NUCLEAR POWER

PROMISE
OR
PERIL?

MICHAEL J. DALEY

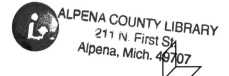
LERNER PUBLICATIONS COMPANY • MINNEAPOLIS

*To Diana Sidebotham and Dr. Judith Johnsrud
who taught me the value of speaking
truth to power.*

Library of Congress Cataloging-in-Publication Data

Daley, Michael J.
    Nuclear power : promise or peril? / Michael J. Daley.
        p. cm. — (Pro/Con)
    Includes bibliographical references and index.
    Summary: Explores opposing viewpoints on expanding the uses of
nuclear power with emphasis on pollution, safety, and waste disposal.
    ISBN 0-8225-2611-5
    1. Nuclear engineering—Juvenile literature. 2. Nuclear energy—
Environmental aspects—Juvenile literature. [1. Nuclear engineering.
2. Nuclear energy—Environmental aspects.] I. Title. II. Series
TK9148.D35 1997
333.792'4—dc20   95-19085

Manufactured in the United States of America
1 2 3 4 5 6 – JR – 02 01 00 99 98 97

CONTENTS

FOREWORD   4

1. PROMISE OR PERIL?   7

2. HOW A NUCLEAR
   POWER PLANT WORKS   21

3. ENERGY AND
   CIVILIZATION   27

4. POLLUTION   45

5. NUCLEAR SAFETY   67

6. WASTE DISPOSAL   99

7. TWO PATHS   117

   RESOURCES TO CONTACT   128

   ENDNOTES   129

   GLOSSARY   135

   BIBLIOGRAPHY   138

   INDEX   141

FOREWORD

If a nation expects to be ignorant and free, . . . it expects what never was and never will be.

Thomas Jefferson

Are you ready to participate in forming the policies of our government? Many issues are very confusing, and it can be difficult to know what to think about them or how to make a decision about them. Sometimes you must gather information about a subject before you can be informed enough to make a decision. Bernard Baruch, a prosperous American financier and an advisor to every president from Woodrow Wilson to Dwight D. Eisenhower, said, "If you can get all the facts, your judgment can be right; if you don't get all the facts, it can't be right."

But gathering information is only one part of the decision-making process. The way you interpret information is influenced by the values you have been taught since infancy—ideas about right and wrong, good and bad. Many of your values are shaped, or at least influenced, by how and where you grow up, by your race, sex, and religion, by how much money your family has. What your parents believe, what they read, and what you read and believe influence your decisions. The values of friends and teachers also affect what you think.

It's always good to listen to the opinions of people around you, but you will often confront contradictory points of view and points of view that are based not on fact, but on myth. John F. Kennedy, the 35th president of the United States, said, "The great enemy of the truth is very often not the lie—deliberate, contrived, and dishonest—but the myth—persistent, persuasive, and unrealistic." Eventually you will have to separate fact from myth and make up your own mind, make your own decisions. Because you are responsible for your decisions, it's important to get as much information as you can. Then your decisions will be the right ones for you.

Making a fair and informed decision can be an exciting process, a chance to examine new ideas and different points of view. You live in a world that changes quickly and sometimes dramatically—a world that offers the opportunity to explore the ever-changing ground between yourself and others. Instead of forming a single, easy, or popular point of view, you might develop a rich and complex vision that offers new alternatives. Explore the many dimensions of an idea. Find kinship among an extensive range of opinions. Only after you've done this should you try to form your own opinions.

After you have formed an opinion about a particular subject, you may believe it is the only right decision. But some people will disagree with you and challenge your beliefs. They are not trying to antagonize you or put you down. They probably believe that they're right as sincerely as you believe you are. Thomas Macaulay, an English historian and author, wrote, "Men are never so likely to settle a question rightly as when they discuss it freely." In a democracy, the free exchange of ideas is not only encouraged, it's vital. Examining and discussing public issues and understanding opposing ideas are desirable and necessary elements of a free nation's ability to govern itself.

The Pro/Con series is designed to explore and examine different points of view on contemporary issues and to help you develop an understanding and appreciation of them. Most importantly, it will help you form your own opinions and make your own honest, informed decision.

Mary Winget
Series Editor

The Vermont Yankee Nuclear Power Plant in Vernon, Vermont, has an impressive safety record and holds a world record in electricity production.

PROMISE OR PERIL?

In the control room of the Vermont Yankee Nuclear Power Plant in Vernon, Vermont, reactor operator Matt is about to throw the switch that will start the production of electricity. Vermont Yankee has been shut down for eight weeks while operators loaded new nuclear fuel into its special steam boiler called a reactor. Unlike the gigantic steam boiler in a coal or oil-fired plant, the boiler at Vermont Yankee is a steel container only 17 feet wide by 63 feet tall. Despite its small size, its single load of 150 tons of nuclear fuel will last for 18 months and produce as much electricity as 2,325,000 tons of coal—more than 23,000 train cars full. And Vermont Yankee will do this without creating any polluting smoke, because the reactor makes heat by nuclear fission—without *burning* anything.

Although the nuclear fuel makes nuclear power plants compact and free of polluting smoke, it presents a unique challenge to the designers and operators of such a plant. During operation, the fuel creates radiation inside the reactor equal to that released by 1,000 atomic bombs—the type of bombs dropped on

Hiroshima and Nagasaki in Japan near the end of World War II. This radiation must be kept inside the nuclear power plant at all times because nuclear radiation can be extremely dangerous—even deadly.

While the new fuel was being loaded into the reactor, the plant's safety systems were inspected, repaired or replaced, tested, and retested. Expert plant engineers as well as government inspectors from the Nuclear Regulatory Commission (NRC) oversaw these operations. Before a nuclear power plant can be started up, everything must conform to rigorous technical guidelines. Matt and all the other reactor operators also spent many hours in training to learn how to handle whatever crisis might occur during operation.

Matt is confident that the start-up will go as planned. But he is also a little nervous. The ability of Matt and all his coworkers to do their jobs correctly protects the lives of the people living near the nuclear plant.

The supervisor gives the word. Matt feels a flutter in his stomach as he throws the switch. His action unlocks the essential force of the universe—the nuclear energy that holds atoms together.

In the reactor, the temperature of the nuclear fuel rises. Pumps as big as a person circulate thousands of gallons of cooling water through the reactor every minute. The constant flow of cooling water is what prevents the nuclear fuel from melting. It carries away the tremendous heat that is used to make steam to run the electric generator. If an accident, such as a broken pipe or human error, should prevent the cooling water from reaching the core for even a few minutes, catastrophe could result. The fuel could melt and release dangerous amounts of radiation into the atmosphere.

But in the unlikely event that something does go wrong, Matt knows that his training, the professionalism of his coworkers, and the many layers of safety systems will keep any problem from escalating into a major disaster. And in the extremely remote chance things did get out of hand, the massive steel and concrete building that surrounds the reactor would prevent any radiation from reaching the public—including Matt's wife at their home in Vernon, and his children, who are in school just across the street from the plant.

During the long, slightly boring hours of monitoring that follow the start-up, Matt thinks about the clean energy Vermont Yankee makes. He thinks of all the oil that is saved—a million gallons per day—and of the oil spills that are avoided.

He wishes that the old, dirty, coal-fired power plants in the Midwest were nuclear plants instead. The pollution from coal contributes to the acid rain problem that is damaging Vermont's forests and poisoning its lakes. Sulfur dioxide from the coal smoke comes down in the rain and causes the soil to release poisons that kill trees. When the rain falls into lakes, it causes the water to become deadly to fish. If the coal plants were nuclear plants, Matt wouldn't have to worry about whether his grandchildren will have maple trees to tap for making syrup, or whether there will be fish to catch in the lakes.

Everything proceeds smoothly and the plant goes "on-line." Vermont Yankee begins to feed 500 million watts of electricity into the power lines of New England—enough to power 500,000 homes.

Operators of nuclear power plants spend many hours in training. Everything they do must conform to rigorous technical guidelines.

Vermont Yankee is one of 110 operating nuclear plants in the United States that supply over 20 percent of the country's electricity.[1] Around the world, 418 nuclear power plants in 25 nations produce about 17 percent of the world's electricity.[2] Advocates of nuclear power think it's a good energy choice for electricity production because it does not pollute the air the way fossil-fuel plants do. For instance, nuclear fission does not produce carbon dioxide (CO_2), one of the gases produced when fossil fuels like coal, oil, and natural gas are burned. Carbon dioxide results when the carbon in the fuel combines with oxygen in the air, which contributes to the greenhouse effect in the earth's atmosphere. Nearly 85 percent of the energy used in the United States comes from fossil fuels. This burning process releases 5.6 billion tons of carbon dioxide into the air each year.[3] Environmental scientists worry that this is causing global warming, which could lead to undesirable climate changes.

Sulfur dioxide (SO_2) is a pollutant released into the air by burning coal. Sulfur dioxide is formed when the sulfur in coal combines with oxygen in the air. Sulfur dioxide goes up the smokestack with the coal smoke. It combines with water vapor in the atmosphere to form sulfuric acid—a part of acid rain and a powerful irritant to human lungs. According to the Union of Concerned Scientists, every ton of sulfur dioxide emitted into the atmosphere results in $3,500 worth of health care costs.[4] In 1992, coal plants emitted over 14 million tons of sulfur dioxide,[5] resulting in health costs of nearly $50 billion. In addition, each year almost 4,000 coal miners die of black lung disease, an illness caused by the inhalation of coal dust.[6]

One of the attractions of nuclear power is that it takes so little fuel to produce so much energy. Fossil fuels, such as petroleum, are in limited supply and often cause pollution.

Petroleum fuels, which provide nearly half the energy used in the world, are produced from crude oil. The earth has only a limited amount of coal and oil, and once that supply is used, people will not be able to produce more. Nevertheless, people are using more and more petroleum each year. If present rates of consumption continue, petroleum may become scarce by the mid-2000s.

Virtually all sources of energy have some drawbacks and side effects, but opponents of nuclear power argue that it is an unacceptable way to make electricity because of its *unique* dangers.

THE PERIL

Ilyich is in the control room of the Chernobyl Nuclear Power Station, Unit 4, in Chernobyl, Ukraine. He is reducing the power of the reactor. The complex operation is almost entirely under automatic control. The process is a little boring. Though Ilyich must monitor hundreds of gauges and meters, there is room in his mind for reflection.

Unit 4 is shutting down for routine maintenance after a long, successful run. The unit is one year old—a showcase of Soviet engineering. The RBMK 1000 design, unique to the former Union of Soviet Socialist Republics (USSR), is believed to be far superior to Western designs. The graphite reactor can be refueled while operating, eliminating the long refueling outages common to U.S. models. Ilyich thinks with pride of the recent *Soviet Life* article featuring his reactor. The article praises it and the Soviet nuclear industry in general for its safety record.

Ilyich also thinks of his nice, large apartment in nearby Pripyat, a benefit of his high-paying, high-prestige job. Soon his shift will end and he will join his family. Perhaps they will pack a picnic and spend the day sunning themselves along the banks of the river. Pripyat, a city of 40,000, was built to house the workers who run the four reactors at Chernobyl. Pripyat is only a few miles from the reactor complex, which is 65 miles northwest of Kiev. Sometimes visitors wonder at that, but the mayor of Pripyat always assures them, "Working in any nuclear power station is safer than driving your car." And the *Soviet Life* article quoted the Minister of Power as saying, "The chances of a meltdown are one in 10,000 years."

It doesn't surprise Ilyich that thoughts of safety come to mind. Before this shift ends, Unit 4 must take part in a small experiment. He and his coworkers have been given an opportunity to contribute even further to the safety of the nuclear industry. They are going to try to squeeze a little extra power out of the turbines as the reactor cools down. If this is possible, it could prove important in an emergency to keep cooling systems running during the few seconds it takes the emergency systems to respond. This extra margin of safety would make the chance of a serious accident even more remote than it is now.

To perform the experiment the workers at the reactor must shut off a number of safety systems. The engineers have reviewed the procedures for weeks and are confident no danger is involved. Still, Ilyich feels anxious when the supervisor finally gives the order to switch off the emergency systems.

On April 26, 1986, a reactor in the Chernobyl nuclear power plant exploded, destroying part of the structure and releasing radioactive fallout across Ukraine, Belarus, and Russia.

Soon Ilyich is too busy to worry. He must keep the reactor running at very low power. It was not designed to do this and now the gauges tell him the fission process is about to stop. If it does, it will take many hours to restart the reactor to continue the experiment. The chief engineer moves to assist Ilyich. Together they manipulate the controls to keep the reaction from dying out completely. They do not realize that while most of the core is going cold, one tiny part of it is out of control. Suddenly, the meters show temperature, pressure, and radiation levels flying off the scale. It is already too late to prevent disaster when Ilyich reaches for the safety shut down.

At 1:23 A.M. on April 26, 1986, the Chernobyl Unit 4 reactor exploded, shattering the containment shell around it. The graphite and uranium fuel burned for several days, creating a cloud of radiation that covered half a continent and eventually drifted around the world.

Ilyich did not survive the massive dose of radiation he received. His family escaped from Pripyat, but not before being exposed to dangerous levels of radiation. They worry about developing cancer sometime in the future. Exposure to high levels of radiation causes a series of reactions in body tissue that results in damage to the body's cells. It can cause disease (such as various types of cancer), injury, or death.

According to a report in the August 1994 *National Geographic,* 150,000 people were evacuated from a 40-mile-wide circle around Chernobyl—an area called the "Zone of Estrangement."[7] No one will be able to live within the zone for decades. According to the Chernobyl Union, a citizens' group in Ukraine, between 1986 and 1994, 5,000 people had died and 30,000 people had become disabled as a result of the explosion. The World Health Organization estimates that over 4.9 million people in Ukraine, Belarus, and Russia were exposed to radiation, but what the consequences will be remains unclear.

Children are especially vulnerable to radiation. Over 300 children have already developed cancers of the thyroid gland, a cancer that used to be extremely rare. The thyroid gland is located in the throat. It regulates metabolism and needs iodine to remain healthy. Radioactive iodine was released in large quantities by the accident. The body cannot tell the difference between regular iodine and its radioactive form, so people's bodies took in radioactive iodine. The radiation is blamed for their cancers.

National Geographic also reports that 30,000 square miles of farmland have been contaminated. Economic costs are in the billions of dollars. Eight years after the Chernobyl explosion, ongoing efforts to deal with the consequences of this one accident were draining 15 percent of the Ukraine government's budget.

THE SCOPE OF THE DEBATE

At the beginning of the nuclear age, advocates of nuclear energy promised the electric companies and the American people that nuclear power plants would provide plentiful electricity that was safe, clean, reliable, and inexpensive. For proponents of nuclear power, Vermont Yankee represents a fulfillment of those promises. Vermont Yankee went into operation in 1972. It has an impressive record of safe operations. The plant has earned citations as the best boiling water reactor in the United States, and it holds a world record in electricity production.[8]

For the people who oppose nuclear power, Chernobyl is the ultimate argument against the use of nuclear power to produce electricity. James Adams, senior analyst for the Safe Energy Communication Council, (SECC), said, "The Chernobyl meltdown was the beginning of the end for commercial nuclear power in the global market. In addition, as more information becomes available on the amount of radiation released worldwide and the continuing environmental and health effects, Chernobyl is truly the accident that never ends."[9]

The SECC, based in Washington, D.C., is an environmental coalition of national energy, environmental, and public interest media groups. With an annual budget of $400,000, SECC works to increase the public's awareness of how conservation and renewable energy sources—such as wind, water, and solar power—can meet an increasing share of our nation's energy needs. Through education and public information campaigns, SECC points out what it believes to be the serious economic and environmental liabilities of nuclear power.

The opponents of nuclear power are a diverse group. Many are ordinary citizens who live near reactors and believe their health and lives are threatened. Others are scientists and doctors. Some nuclear opponents belong to environmental groups such as Earth First! and Greenpeace. While these individuals and groups do not speak with a unified voice on all issues, they are united in the belief that nuclear power has not delivered on its promise to provide safe, inexpensive power, and it has presented society with the possibly insoluble problem of nuclear waste.

Not only special interest groups of antinuclear activists oppose nuclear power. SECC points to polls showing that 62 percent of Americans do not want new nuclear power plants to be built. Indeed, SECC suggests that even electric utilities lack confidence in nuclear power. Every nuclear plant ordered after 1973 has been canceled, and no new plants are likely to be ordered soon. SECC executive director Scott Denman

A Greenpeace activist carries wooden crosses to be set up near a nuclear power plant (in background) *in Czechoslovakia to commemorate Chernobyl.*

said, "Nuclear power is a relic of a failed energy policy. It's time to shift to sustainable, affordable energy options that are available today."[10] SECC believes that existing nuclear power plants should be phased out of operation as soon as possible, and future supplies of electricity should come from renewable sources.

These views are not shared by proponents of nuclear power, such as the Nuclear Energy Institute (NEI). Nuclear power supporters think that Chernobyl is a disaster that cannot happen in the United States. NEI spokesperson Scott Peters says, "The Chernobyl-style plant has never been built outside the Soviet Union and could not be licensed to operate in any other country. The technology is simply not up to western standards. For instance, unlike Chernobyl, western-style reactors all have massive concrete and steel containment structures. Our reactors also shut themselves off if they overheat instead of flashing out of control as the Russian reactor can."[11]

The NEI is a private, nonprofit association of nearly 400 companies, located around the world, that have an interest in nuclear power. NEI includes electric utilities, reactor manufacturers, labor unions, universities, and research institutions. Based in Washington, D.C., its mission is to build understanding of and support for nuclear energy. NEI points to polls that contradict the surveys of SECC. Their polls show that in 1991, 73 percent of Americans felt nuclear energy should play an important role in meeting our energy needs.

NEI blames the lack of orders for new plants on unnecessary and burdensome regulation by the NRC and on the disruptive activities of people opposed to nuclear power. NEI believes the public has a false impression about the risks of nuclear accidents and radiation. Antinuclear activists prey on these unreasonable fears. For example, there are people who say "not in my backyard" (NIMBY) to any attempt to find sites for nuclear waste disposal facilities. In NEI's view, the technology for safe disposal of nuclear waste—burial in rocks far below the earth's surface—has long been available. Only misperceptions of the danger lead the public to oppose this safe solution to the problem.

Scott Peters says, "We've been safely handling nuclear waste for more than 50 years. We have the waste from nuclear weapons programs, from generating electricity, from nuclear medicine, and many other uses of the technology, and we must dispose of it responsibly."[12]

With a budget of almost $18 million per year, NEI produces information to counteract the public's misunderstanding of nuclear technology. According to NEI, nuclear power has a vital role to play in meeting the world's growing needs for electricity because it does not contribute to the threat of global warming or deplete fossil fuel supplies. NEI argues that we need to keep the existing nuclear plants operating and extend their licenses for another 20 years. Furthermore, according to NEI, we must start building new nuclear power plants to meet the energy needs of the future.

SECC and NEI are not the only groups for or against nuclear power, but these two groups work on a national scale to influence the American public's perception of nuclear power. Their arguments and supporting information are of special interest in the debate about nuclear power.

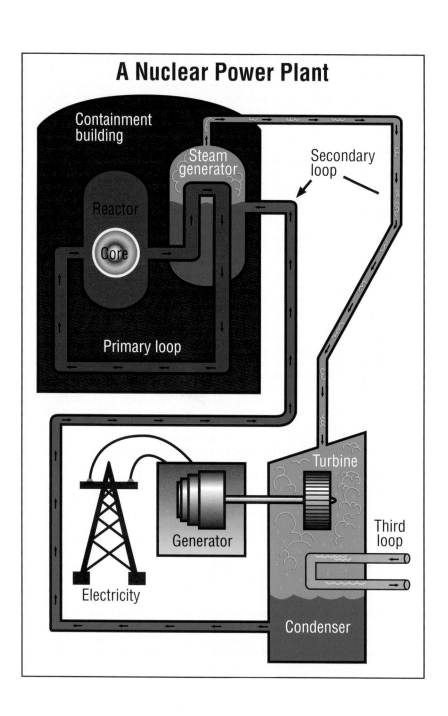

HOW A NUCLEAR POWER PLANT WORKS

A nuclear power plant makes electricity so that the electric company can meet its customers' needs for power and earn money for the people who own the power plant. In this way, nuclear power plants are like other kinds of power plants. What makes them unique is the fuel. In a coal-burning plant, the furnace burns coal to make the steam that turns the generator to produce electricity. A nuclear power plant also makes steam, but its nuclear furnace—called a reactor—has no flame. The heat is generated by nuclear energy released in the process of fission, or splitting. One ton of nuclear fuel produces as much energy as 15,500 tons of coal. And all the nuclear fuel needed to produce electricity for 12 to18 months fits inside the reactor, which is a steel container 17 feet wide by 60 feet tall.

Nuclear fuel is made from a type of uranium called uranium 235 (U-235). Uranium is one of 109 known chemical elements. All matter is made up of chemical elements. Elements, in turn, are composed of atoms.

All atoms have protons and neutrons at their center and electrons that orbit the center. In some atoms, the arrangement of protons and neutrons is unstable, and the atom may break apart or decay. Atoms like these are called radioactive. Scientists learned that the natural process of radioactive decay can be duplicated by striking an atom of U-235 with a neutron. The atom becomes unstable and splits (fissions) into two smaller atoms—called fission products—that may also be radioactive. The atom also releases more neutrons and huge amounts of heat energy. If there are enough other U-235 atoms concentrated nearby, the neutrons will hit them, and they will also fission, releasing more neutrons and more heat. This chain reaction is like the effect you get if you line up a long row of dominoes, then topple the first one. The chain reaction will continue until there is no more U-235 fuel.

The heart of a reactor is the core, where all the heat is generated. The core contains the fuel rods. These are 12-foot-long metal tubes filled with ceramic pellets of U-235 that are about one-half inch wide. The fuel rods are bundled together to form fuel assemblies, which are inserted into the reactor core along with control rods. The control rods are made of a material that absorbs neutrons. By moving the control rods into or out of spaces between the fuel assemblies, the operators of the reactor can regulate how fast the fission process takes place. When the control rods are all the way in, no fission can occur because the rods absorb so many neutrons that a chain reaction cannot begin. Once they are removed, the core begins to heat up. The heat is carried away by a coolant, which is pumped through

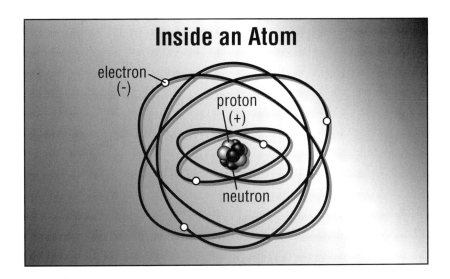

Inside an Atom

electron (-)

proton (+)

neutron

spaces between the fuel assemblies. The coolant can be water, a gas such as helium, or liquid metals such as sodium. Water is used as a coolant in all commercial U.S. reactors.

Two types of reactors produce electricity in the United States: boiling water reactors (BWR) and pressurized water reactors (PWR). In a BWR, the water boils to steam inside the reactor. The steam then goes from the reactor directly to the generator in what is known as a single loop. In a PWR, the water is kept under such high pressure in the reactor that it cannot boil. Instead, this superheated water is circulated through another water-filled container called a steam generator. The steam made in a PWR's second loop is what goes to the generator.

In both designs, the water that contacts the nuclear fuel is kept in a closed loop that constantly recirculates back into the reactor. But the steam coming out of the

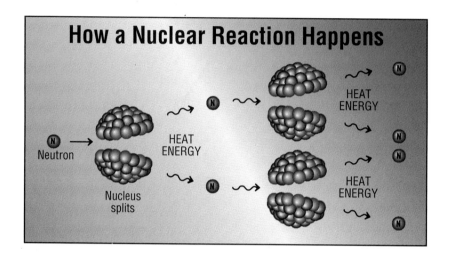

How a Nuclear Reaction Happens

Neutron

HEAT ENERGY

Nucleus splits

HEAT ENERGY

HEAT ENERGY

generator cannot be pumped into the reactor. It must first be condensed to liquid form, using cold water. This water comes from lakes and rivers. The large amount of river or lake water used by a nuclear power plant never (under normal conditions) comes into direct contact with the water in the reactor's closed loops, so rivers aren't contaminated by radioactivity.

As part of normal operations, some of the radioactive fission products seep out of tiny cracks that form in the fuel rods. These radioactive products get into the coolant and must be removed. Some of the fission products are radioactive gases, which are vented into the atmosphere. Some of them are solids and are released into nearby waters or collected for disposal as low-level nuclear waste. Every 12 to 18 months, about one-third of the fuel (30 to 50 tons) must be replaced because the U-235 has been used up.

The used fuel rods are called spent fuel. Spent fuel is extremely radioactive because the rods contain hun-

dreds of radioactive fission products, such as plutonium, cesium, and strontium. When spent fuel is removed from a reactor, it must be handled under 30 feet
of water. Water is a moderator. It slows neutrons and,
thereby, the nuclear reaction. The water shields workers from the spent fuel's intense radiation. If it were
ever accidentally lifted out of the water, the exposure
would quickly kill anyone standing nearby. Spent fuel
is also called high-level nuclear waste. All the nuclear
fuel ever used in the United States is stored in pools of
water at the nuclear plants.

The most critical thing about operating a reactor is
preventing the fuel rods from becoming hot enough to
melt. Melting would release the enormous amounts of
radiation locked inside them. The fuel must constantly
be covered by the coolant to keep it from melting.

Reactors must be able to survive a loss-of-coolant accident (LOCA), which would most likely occur because of a broken pipe. In emergencies, the control
rods must insert instantly to stop the chain reaction.
Even with the fission stopped, heat is still generated by
the fuel. The emergency core cooling systems (ECCS)
must begin working within 15 to 30 seconds after a
leak occurs to remove the heat and supply new
coolant. The ECCS may not prevent fuel damage in all
accidents, so most reactors have a containment building, a large building that completely encloses the reactor. The containment is lined with steel, and its
concrete walls are four feet thick. It is designed to
withstand the enormous heat and pressure of an accident and to prevent radiation from reaching the outside environment.

During most of human history, human and animal muscle was the main source of energy.

ENERGY AND CIVILIZATION

What makes human beings unique? The opposable thumb? The ability to walk upright? A highly developed brain? Language? Or is it our ability to use forms of energy other than that available from our muscles?

Some animals can cooperate with the natural forces around them. The hawk, for example, can soar for hours on a thermal (a rising body of warm air) without once flapping its wings. But human beings *seize* control of energy. Instead of depending on a thermal, we drive a propeller with an engine and take our airplane in any direction we desire. We use energy to overcome the limitations of night, to blast tunnels through mountains, to reach across space to the moon, to transform the elements into new substances and make them obey the designs of our minds.

Modern people associate energy with fuels and machines, but for the vast majority of human history, the primary energy came from human and animal muscles, fueled with food grown with energy from the sun. There was also wind power, water power, and fire. Physicist Glen Seaborg has pointed out that ancient civilizations relied on the exploitation of many people to achieve a high standard of living for the few: "Classical Greece—with its Aristotle, Plato, and Homer—was built by a population of some 34,000 free men supported by a foundation of about 300,000 40-watt machines; that is, 300,000 human slaves. The 'glory' of ancient Rome depended on 15 to 20 million citizens who commanded about 130 million slaves—still generating 40 watts of mechanical power apiece."[1]

It wasn't until the invention of the steam engine that people acquired a new slave, one that was tireless and more powerful than a hundred human workers. All it required for ceaseless activity was a steady supply of fuel—first wood, then coal, and finally oil.

In the late 1940s, the Atomic Energy Commission asked engineering consultant Palmer Putnam to estimate the amount of energy used throughout history.[2] Putnam determined that until 1860, the human race had managed to use up about 7 Q of energy. (Q stands for quad, a billion billion British thermal units.) One quad will meet the energy needs of 3 million average Americans for one year.[3] When Putnam added everything up between 1860 and 1947, he discovered that people had used 12 Q in those 87 years—almost twice as much energy as all the energy used in the previous 7000 years![4] That level of energy consumption could be compared to having about 7 slaves working for each person, still pretty much what the Roman citizen enjoyed. [5]

In the United States, we use about 88 Q per year—over 4 times the amount used throughout human history. This is like having 150 slaves working for each person, every day, all day.[6] Worldwide, humanity consumes 350 Q each year.

All that energy helps support our standard of living. Think about the ways you used energy this morning between the time you woke up and arrived at school. Did your electric alarm clock go off? Did you turn the lights on? Take a shower? Cook breakfast? Maybe you just grabbed a glass of orange juice from the refrigerator? Was the house warm? Did you listen to the radio or watch TV? These examples are just the tip of the iceberg, however. Residential energy use accounts for less than one-third of all energy consumption.

If you flushed the toilet or threw away some trash, you used energy you can't see—energy at the sewage plant and the landfill. Did you take a bus or car to school? Don't forget the energy used to produce the orange juice in your refrigerator or the bread you popped in the toaster. This hidden energy is used by the industrial and transportation sectors of our economy.

Electricity is the fastest-growing form of energy. It enables us to use many time-saving appliances that we have come to rely on.

You might have noticed the predominance of things run with electricity in your home. Though the charts show electricity is only 30 percent of the total amount of energy consumed, according to the Nuclear Energy Institute, it is the form of energy for which the demand is growing the fastest. In 1995 Americans used 72 percent more electricity than they did in 1973.[7]

WHEN DEMAND OVERTAKES SUPPLY

Energy is the lifeblood of modern civilization. Worldwide, over 14 million tons of coal, 66 million barrels of oil, 191 billion cubic feet of natural gas, and 98 million million watts of electricity are used every day all year round.[8] At these rates of consumption, the known supply of oil will run out in 40 years. The supply of coal will run out in 700 years, and natural gas in 60 years.[9] Yet for all the energy human beings use, many people are still without the improved health, long lives, and material comforts we enjoy in the United States. Only eight percent of the world's people have automobiles.[10] Half of the world's people—over 2.5 billion—do not have commercial energy for factories, businesses, schools and hospitals. The same number have no electricity in their homes. Muscle power and wood remain their primary energy sources.[11]

Meanwhile, to support our standard of living, the average person in North America uses 20 times more energy than a person living in South Asia.[12] With only 4.6 percent of the world's population, the United States consumes 24 percent of the world's energy.[13]

As the rest of the world aspires to improve their lives, energy consumption rates will increase. The

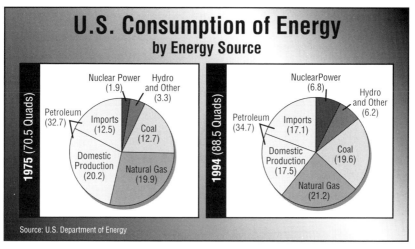

U.S. Consumption of Energy
by Energy Source

1975 (70.5 Quads)

Nuclear Power (1.9)
Hydro and Other (3.3)
Petroleum (32.7)
Imports (12.5)
Coal (12.7)
Domestic Production (20.2)
Natural Gas (19.9)

1994 (88.5 Quads)

NuclearPower (6.8)
Hydro and Other (6.2)
Petroleum (34.7)
Imports (17.1)
Coal (19.6)
Domestic Production (17.5)
Natural Gas (21.2)

Source: U.S. Department of Energy

The U.S. primary energy demand increased by more than one-fourth between 1975 and 1994.

amount of energy used in the world in the 1990s is 66 percent greater than the amount used in 1960.[14] It will increase another 75 percent by the year 2020.[15] If the fossil fuels we rely on so heavily are running out and causing such undesirable environmental harm, where will new power come from?

On August 6, 1945, the answer seemed clear.

THE ATOMIC AGE BEGINS

The atomic age was the result of a 1939 experiment by two Austrian physicists. Otto Hahn and Fritz Strassman proved that uranium fissioned when hit by neutrons. Their work was the culmination of work by pioneering scientists such as Marie Curie and Albert Einstein. It set the world of physics afire.

Boyd Norton, a reactor physicist, said: "The process of nuclear fission is awesome, beautiful, elemental, and elegant; it gets down to the very roots of the universe. The discovery, exploration, and utilization of nuclear fission is one of mankind's greatest intellectual achievements."[16] Those words do not represent the view of just one scientist. They express the feelings of hundreds of scientists who participated in bringing this new energy form into the world.

In the early 1900s, the field of atomic energy was brand new. It was one of the most exciting areas of physics for young scientists to work in. Atomic scientists progressed from the discovery of fission in 1939 to the first sustained chain reaction in only three years. That event occurred in 1942, when Enrico Fermi built the first atomic *pile,* or reactor, in a squash court at the University of Chicago. Three years later, the first atomic bomb exploded. Norton compared this feat to achieving a lunar landing only six years after the launch of the first crude rocket—an effort that actually spanned 40 years.

J. Robert Oppenheimer, an American physicist, was the head scientist of the top-secret Manhattan Project, which produced the atomic bomb. The United States was anxious to develop a nuclear bomb because it had entered World War II in 1941, and it wanted to be the first country to have such a powerful weapon. A test bomb was first exploded in the desert near Alamogordo, New Mexico, on July 16, 1945. Temperatures at the center of the blast almost equaled those at the sun's center. On witnessing the fireball produced by the test, Oppenheimer said, "I am become Death, the Shatterer of Worlds," which is from a passage in the Bhagavad Gita, the sacred book of the Hindus.[17]

Enrico Fermi, left, *won the 1938 Nobel Prize for physics for his work on nuclear processes. Robert Oppenheimer*, right, *became known as the father of the atomic bomb.*

Reporter William Laurence witnessed the test and offered this interpretation of the meaning of the mighty shockwaves that boomed over the desert: "The hills said 'yes' and the mountains chimed in 'yes.' It was as if the earth had spoken and the suddenly iridescent clouds and sky had joined in one mighty affirmative answer. Atomic energy—yes."[18]

Other observers were less given to poetic interpretation. They recorded that the earth near the test site had been turned into a radioactive green glass and that the herds of native antelope disappeared forever from that part of the desert.[19]

For the world at large, the atomic age began on Monday, August 6, 1945, when the *Enola Gay,* a U.S. B-29, flew over the Japanese city of Hiroshima and dropped an atomic bomb. On Thursday, August 9, a second atomic bomb devastated Nagasaki. These were acts of such dramatic destruction that "the whole world gasped" and "could talk of little else," for "with this newly released force, man can destroy himself or create a world rich and prosperous beyond all previous dreams. From the day of the alchemists, generations of scientists have tried to unlock this energy. Now, when the fantastic dream has literally and figuratively burst into reality, men and women find themselves shaken and wondering whence it came, what it is, what it will mean in our daily lives—and how we can control it."

Those words appeared on the cover of *The Atomic Age Opens,* an "instant book," published the same month the atomic bomb was dropped. The book contained excerpts from news articles from around the world. The majority expressed praise for the end of World War II and awe at the achievements of science. Some expressed fear, some horror, but all accounts were tinged with a recognition that a new, unique chapter of human history was unfolding.

The Manhattan Project was a cooperative effort among the military, the government, and the scientific community. With the help of 65,000 workers, they

Hiroshima, Japan, after the atomic bomb exploded

built the largest industrial complex in the world from scratch. The effort cost $2 billion. Scientists came from all over the free world.

After so much effort and scientific brilliance had been devoted to destruction, many wondered: Could atomic power also be used for the good of people? In 1945, dread and hope intermingled. Dr. R. M. Langer, a physicist, envisioned a world completely altered by the use of U-235. "The power behind the atomic bomb can be harnessed to produce the Utopia that men have dreamed of through centuries of war, depression, famine, and disease.... U-235 could bring within a century a world in which there is no need for war, a world in which there are no houses, railroads, or highways."[20]

Dr. Langer believed U-235 would bring untold wealth to all humanity. No one would have to work. People would live underground in houses lit by the fluorescent glow of U-235, eat food grown in chambers filled with U-235, drive cars powered by a tiny chunk of uranium, and fly in airplanes propelled by the thrust of U-235's radiation.

While such visions filled the popular press, hundreds of the atomic scientists who had built the first

bomb were organizing to save civilization. They knew
the soon-to-be-developed hydrogen bomb meant nuclear war could never be fought without destroying the
entire world. Soon, they called for the international
control of atomic weapons to avoid a dangerous arms
race.

These same scientists believed that if civilization
could survive the threat of nuclear annihilation, the
atom held enormous potential for good in medicine,
agriculture, and industry. The public was encouraged
to dream along with the scientists about the fantastic
atomic future, and children were shown movies like
Our Friend the Atom. The challenging work of turning
these dreams into reality became the job of the U.S.
Atomic Energy Commission.

MAKE A BLESSING OF IT

The Atomic Energy Act was passed by Congress and
signed into law by President Harry S. Truman on August 1, 1946. The act established a five-member Atomic
Energy Commission (AEC) that would have control
over both military and commercial development of
atomic energy. The government would own all the materials and facilities, and it would keep much of the
knowledge secret. The act's author, Senator Brien
McMahon of Connecticut, said, "Never before...has
Congress established an administrative agency vested
with such sweeping authority and entrusted with such
portentous responsibilities."[21]

The AEC's responsibility, besides building atomic
weapons, was to direct the use of atomic energy "toward improving the public welfare, increasing the

President Harry S. Truman signed the Atomic Energy Act, a bill designed to safeguard information and develop ways to use atomic power for world peace.

standard of living, strengthening free competition... and cementing world peace."[22]

David Lilienthal, the first chairperson of the AEC, remembers that the early years were devoted to building bombs. Everyone, even President Truman, assumed the United States had a stockpile. But at his first meeting with Truman, on April 3, 1947, Lilienthal had to tell the president that, "There was not a single operable bomb in the 'vault' at the Los Alamos atomic arsenal. Nor could there be for many months to come."[23]

At the end of that meeting, Truman told Lilienthal, "You must make a blessing of it [atomic energy] or [he pointed to a large globe] we'll blow that all to smithereens."[24] According to Lilienthal, the "blessing" would be "to convert the virtually unlimited power of nuclear fission to electricity."[25] But it would be six years before another president would try to make the blessing a reality. Speaking before the United Nations General Assembly on December 8, 1953, President Dwight D. Eisenhower launched the Atoms for Peace program. He said, "The United States pledges.... to devote its entire heart and mind to find the way by which the miraculous inventiveness of man shall not be dedicated to his death, but consecrated to his life."[26]

This ambition to extend the peaceful uses of the atom to the whole world came in the midst of a desperate arms race and a cold war with the Soviet Union. In the same speech, Eisenhower assured any enemy that the United States would lay waste to its country even as it laid waste to ours—an expression of the policy known as MAD: Mutual Assured Destruction. It was a time when children might watch *Our Friend the Atom* one day and practice "duck and cover" routines in case of atomic attack the next day. The world was just learning that radiation was a silent, secret killer that stretched its hand out over time—the first, unexpected wave of cancers emerged in the survivors of Hiroshima and Nagasaki. Scientists alerted the public to strontium, a radioactive isotope present in fallout from nuclear explosions. Strontium can be absorbed in the bones of humans and animals. It began to appear in the milk that children drank every day. Eventually this sparked worldwide protest against bomb tests in the atmosphere.

The newly emerging nuclear industry accomplished great technical triumphs during the 1950s. The first atomic submarine, the U.S.S. *Nautilus,* was launched in 1954 and cruised 1300 miles underwater

Schoolchildren stop their lessons to practice a "duck and cover" routine in case of an atomic bomb attack.

on its maiden voyage.[27] Arco, Idaho, became the first town powered by atomic energy on July 17, 1955. The BORAX III reactor, used to light Arco, was an early prototype of today's boiling water reactors such as Vermont Yankee.[28] BORAX III was located at the Argonne National Laboratory, near Arco.

But many other glorious visions were revised by reality. The industry abandoned development of atomic airplanes. What if one of them crashed? The use of atomic bombs for canal digging was abandoned because fallout was too dangerous. The development of atomic cars was abandoned because passengers would need 50 tons of lead shielding.

In 1957, ten years after Truman's plea for a peaceful use of atomic energy, the Shippingport, Pennsylvania, reactor went into operation. Shippingport was a larger version of a submarine reactor like the one that powered the *Nautilus*. It was built as a demonstration project by the Atomic Energy Commission. It supplied electricity to the Pittsburgh area until 1982. Shippingport's successful operation demonstrated that nuclear technology was ready to produce electricity for U.S. homes and industry. However, its tiny power rating of 60 megawatts (1 megawatt = one million watts) was not large enough to attract serious interest from electric power companies. Power companies were used to building plants of 500 megawatts or more. In order to make nuclear power useful to power companies, reactor engineers had to overcome immense technical obstacles to design 200-, 500-, and finally 1000-megawatt reactors.

Boyd Norton began his career as a physicist testing reactors at the National Reactor Testing Station in Idaho Falls, Idaho. He noted that some scientists in the Atomic Energy Commission worried that the drive for large reactors had gotten ahead of the research needed for basic safety provisions. For example, they hadn't proved that the all-important emergency core cooling system would work as planned.

Utilities were skeptical of the high costs of nuclear power compared to the use of cheap coal. They also worried about liability (lawsuits) if an accident should happen. No private insurance company would take on the unknown risks of a nuclear accident. To resolve this, Congress passed the Price-Anderson Act in 1957. It provided that in the event of an accident, no one could sue the utility or manufacturer for damages, even in cases of carelessness or negligence. The Price-Anderson Act also set up a small fund of $560 million

 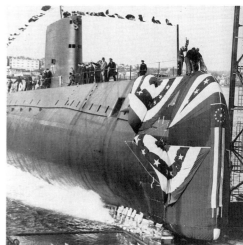

Arco, Idaho, the first community to be lighted exclusively by nuclear power (left) and the Nautilus (right), the first nuclear-powered submarine

to compensate the public for damages caused by a nuclear accident. Industry critics considered this an unfair subsidy. No industry before or since has ever enjoyed such protection from liability. Critics maintain that the funds are too low to really help the large numbers of people who might be hurt or have their property ruined forever.

Though intended to be just helpful start-up legislation, Price-Anderson still remains in force. Defenders of the nuclear industry point out that the fund is now paid for by money from nuclear utilities and contains $7 billion. It remains a highly controversial law, however, especially with Chernobyl cleanup costs running into hundreds of billions of dollars.

Even with the liability problem solved, Shippingport produced electricity at 12 times the cost of coal, so utilities hesitated. But most new technologies are expensive at first. So reactor manufacturers took a gamble on the future and offered reactors for sale below cost to jump-start the industry.[29] In 1964, Jersey Central Power and Light ordered a 650-megawatt reactor from General Electric. The plant was called Oyster Creek, and it represented the first nuclear power plant justified—from the utilities' point of view—on purely economic grounds. This "loss-leader" deal cost General Electric about $30 million, but it did begin the era of commercial nuclear power.[30] Orders flowed in, and many people believed that nuclear energy was finally ready to fulfill the promise of clean, safe electricity that would be almost "too cheap to meter."

But by 1970, nine years before the accident at the Three Mile Island nuclear plant near Harrisburg, Pennsylvania, where workers mistakenly shut off the emergency cooling system, and long before the Chernobyl accident, the vision of a nuclear-powered America was in trouble. The last order for a reactor was placed in 1974. Scientists and engineers began to speak out about the dangers of radiation and reactor safety. Construction costs began to increase, and citizen opposition grew.

Despite falling far short of the 1,000 reactors the AEC envisioned, 110 operating plants do supply 20 percent of electricity in the United States. Energy demand continues to grow while fossil fuel supplies are being depleted, and urgent environmental questions are being raised—especially about the threat of worldwide climate change from global warming. Proponents therefore believe that nuclear power, which does not produce carbon dioxide, could help prevent global warming.

Supporters insist there is an urgent need to recommit the public to the vision of a nuclear-powered future. One is Chauncy Starr, founding president of the Electric Power Research Institute. When asked by *Nuclear Industry* magazine to look ahead, Starr said, "If you're a true environmentalist...you have to be a pro-nuclear enthusiast."[31] Because developing countries do not have the money or technical expertise to use nuclear energy, Starr believes it is the "ethical responsibility" of industrialized countries to use it and leave the fossil fuels to developing nations.

But should we continue on a path that brought the disasters at Three Mile Island and Chernobyl and created an unsolved nuclear waste problem? Do we actually have a choice?

As we examine the arguments of advocates and opponents, keep these words of David Lilienthal in mind:

> Why does the ordinary citizen need to know anything at all about atomic energy? Why can't one leave the entire business...[to the experts]... whose job it has been...to design and build and operate nuclear plants?
>
> The answer is because these experts are no more infallible than any other experts—which is to say they are very fallible indeed—and they need to be subjected to the checks and balances of cross-examination and inquiry that are at the heart of the democratic process in an open society.[32]

Pilgrim Nuclear Power Station in Plymouth, Massachusetts

POLLUTION

On December 30, 1988, operators in the control room of the Pilgrim Nuclear Power Station in Plymouth, Massachusetts, began to allow the nuclear fuel in the reactor to heat up. It would soon be generating electricity for the Boston area. Outside the gates, about 125 local citizens gathered to protest the start-up. Carrying signs that read "Close Pilgrim's Cancer Factory," 27 protesters deliberately trespassed onto the plant grounds and were arrested. They felt compelled to commit this act of civil disobedience because, as a normal consequence of its operation, they thought Pilgrim would begin releasing radiation into the environment. Pilgrim and many other boiling water reactors release radioactive gases from a tall "smokestack." This process allows the concentrated gases to mix with the air before people inhale them. The protesters feared this radiation would cause cancer.

"This all started at the grassroots with neighbors worried about neighbors getting sick," said Mary Ott of Citizens Urging Responsible Energy. People went door-to-door to gather information about local illnesses.

A "smokestack" at the Vermont Yankee plant can be seen at the far right.

Their work convinced Massachusetts state officials to conduct a formal health study. The study revealed 50 percent more cancer cases in the five towns near the Pilgrim plant than in the rest of the state.[1]

Boston Edison Company official Ralph Bird asserted, "There has not been enough radiation released in its [Pilgrim's] lifetime to cause a single cancer in the public." And he accused state health officials of creating "unnecessary public anxiety" by suggesting there might be a link between the cancers and Pilgrim. "There is no risk to the public in operating this plant. If there were, we wouldn't operate it. Period."[2]

Bird claimed that a person living in a tent near the fence around Pilgrim would receive an annual dose of radiation from the plant that was 400 times less than they would receive from natural sources of radiation, such as sunlight or cosmic rays. When experts estimate the dose that Bird is talking about, they assume the radioactive gases have been spread evenly in a huge circular area around the plant. Thus, a single person is likely to breathe in only a tiny amount of the radioactive gas, if any at all.

Epidemiologist Sidney Cobb suggested this model might not be true for Pilgrim. An epidemiologist studies patterns of illness in large groups of people and tries to find the cause. This is not an easy task, since people are exposed to hundreds of germs, viruses, and dangerous pollutants every day. Pilgrim is located on the coast where the prevailing ocean breezes carry the air over the same few towns most of the time. Cobb wondered if the radioactive gases weren't also being carried by these breezes. If that were true, the people living in those towns would receive a much higher dose than experts calculated when using Pilgrim's model. Working with Cobb's ideas, a Massachusetts state-funded study showed a strong association between radiation releases from Pilgrim and an increased risk of getting cancer—up to four times higher for people who lived within 10 miles of the plant than for people who lived farther away.[3]

At about the same time, however, a study released by the National Institutes of Health (NIH) concluded that there was no evidence of increased cancer risk to people living in counties with, or adjacent to, nuclear power plants, including the Pilgrim plant.[4]

The dispute about health effects from exposure to low levels of radiation is not unique to Pilgrim—or to

Accidents such as this gas plant explosion in Mexico City demonstrate that all energy sources carry risks to human health and the environment.

nuclear power plants alone. It has occurred near facilities associated with every aspect of the nuclear fuel cycle. The concerns include excess lung cancer in uranium miners, leukemia in the children of workers at British reprocessing plants, illnesses among U.S. soldiers exposed during A-bomb tests, and excess cancers among workers at enrichment plants (where nuclear fuel is made) and on nuclear submarines. As in the Pilgrim controversy, scientific studies are available to support both sides of the debate.

Taken in isolation, the existence of so much controversy can be disturbing. But other energy sources also have adverse impacts: coal miners develop black lung disease, or are killed in cave-ins; air pollution from power plants using fossil fuels increases respiratory illnesses; and gas pipelines explode.

The question, say advocates of nuclear energy, is not whether radiation involves risks to human health, but whether it is riskier than the other dangers we face.

POLLUTION IN OUR WORLD

Pollution is a reality of modern civilization. Pollutants are in the air we breathe, the water we drink, and the food we eat. The Environmental Protection Agency (EPA) monitors the discharge of thousands of different substances into the environment. Study of these chemicals is fairly new, and the effects of many of these pollutants on human health are not well understood. In contrast, the first standards for protecting people from radiation went into effect more than 60 years ago. The NEI says that more is known about the health effects of radiation than about most other physical or chemical agents.[5] This knowledge gives them the confidence to say that no member of the American public has ever been harmed by exposure to radiation from a nuclear power plant.

Opponents counter that the industry can make such claims because leukemia caused by radiation is just the same as leukemia that developed from other causes. Even genetic disorders may contribute to the disease. These factors make it very difficult to demonstrate a link between radiation released into the environment and cancer. Kay Drey is a board member of the Nuclear Information and Resource Service (NIRS), an anti-nuclear organization based in Washington, D.C. Drey argues that, because "any exposure to radiation increases the risk" of mutation, cancer, or disease, "no one has the right to add man-made radiation to the environment."[6] The SECC notes that "since 1928, the

This solar-powered lighthouse on Gaspé Peninsula in Quebec, Canada, provides an example of practical alternative energy use.

radiation standards have been made increasingly more restrictive as more is learned about the effects of radiation."[7] SECC believes that the public does not have to accept the risk from nuclear pollution at all. In SECC's view, much safer energy alternatives exist, such as conservation, and renewable energy sources such as wind power, water power, and solar power. These alternative sources produce no air pollution, nuclear waste, or radiation.

Nuclear proponents emphasize the need for perspective when weighing the risks from radiation. They must be compared to other risks society tolerates to enjoy the benefits of modern technology—such as the release of carbon monoxide, hydrocarbons, and nitrogen oxides into the air by automobiles. Automobile accidents kill approximately 50,000 people each year. Such risks are part of contemporary life.

Nuclear proponents insist there are risks involved in SECC's ideas. They believe conservation and renewable energy cannot meet the needs for electricity that help maintain our high standard of living. NEI Vice President John Siegel calls energy conservation "a pseudonym for electricity shortages," while NEI spokesman Scott Peters thinks our nation has "already squeezed about every drop we can out of efficiency."[8]

NEI believes our need for electricity is so great that we must exploit every energy source available to us. In its view, if we eliminate the nuclear option, we will have to rely even more on coal. Nuclear proponents often compare the risks from nuclear plants to the risks from coal plants. They ask, "Would the citizens who protested against Pilgrim be better off if the reactor were a coal furnace?" Some research indicates that pollution from each coal plant causes between zero and eight deaths each year, as well as lung disease.[9] In his book arguing for the use of nuclear power, *Before It's Too Late,* scientist Bernard Cohen attributed about 5,000 deaths each year to coal used for electric power. These deaths result from people breathing in polluted air. The effects of pollution can be significant during periods when weather conditions trap high levels of smog in certain areas. For example, Cohen writes, "There were at least eight episodes in London between 1948 and 1952 in each of which hundreds of excess deaths were recorded, the largest in December 1952 when 3500 died. There were three episodes in New York City...November 1953 causing 360 deaths, January-February 1963 leading to 500, and November 1966 responsible for 160."[10] Cohen argues that even if we accept (and he does not) the antinuclear people's claim that radiation releases from nuclear fuel kill about 100 people each year, it is clear that nuclear power remains a far better choice than coal.

The Committee for Nuclear Responsibility, an antinuclear group in San Francisco, California, objects to this kind of reasoning. It amounts to asking: How many people are you willing to let die so you can have electricity?[11] The committee believes that society should be seeking electricity supply technologies that do not force us to make such grim choices.

For many people opposed to nuclear power, the comparison between nuclear energy and coal remains a false one. SECC points to a report that shows energy efficiency alone could replace Pilgrim: "In New England, the region of the country most dependent on nuclear and oil-fired electricity, the New England Policy Council estimates that a 35 percent reduction in demand is possible using currently available technology. Further advances [in efficiency technology] could raise that figure to 57 percent, saving the equivalent output of as many as 15 nuclear plants."[12] There are only eight nuclear plants in New England.

THE ENVIRONMENTAL BENEFITS OF NUCLEAR POWER

Human health is not the only concern when considering the pollution problem. Environmental impacts, such as acid-rain damage to lakes and forests or the threat of global warming caused by carbon dioxide, must also be considered. Proponents contend that nuclear power is vital in combating such environmental threats and offering an alternative to our dependence on the limited supplies of fossil fuels. Before the widespread use of nuclear power, most electricity came from burning fossil fuels. About 15 percent came from burning oil, a situation that made the United States dependent on foreign sources. Burning coal, oil, and natural gas contributes to three major air pollution problems: carbon dioxide (CO_2), which contributes to global warming; sulfur dioxide (SO_2), a contributor to acid rain, which poisons lakes and destroys forests; and nitrous oxide (N_2O), a major part of smog as well as a component of acid rain.

According to NEI, between 1973 and 1993, nuclear power's share of the U.S. electricity supply grew from 4 percent to 19 percent, while the use of oil to produce electricity declined from 15 percent to only 4 percent.[13] Proponents claim credit for displacing the use of oil

Drivers wait in line for gas in New York City. On this June day, 90 percent of all stations were closed because of gas shortages during the oil embargo.

and assert that our air quality would be much worse without the increased use of nuclear power. By substituting for the use of fossil fuels in electricity production over those 20 years, nuclear power replaced the burning of 2.1 billion barrels of oil, 2.6 billion tons of coal, and 9.0 trillion cubic feet of natural gas with energy from 30,000 tons of uranium. This prevented the emission of 1.6 billion metric tons of CO_2, 27 million tons of N_2O, and 65 million tons of SO_2. Nuclear power has also increased our energy security by reducing our need for foreign oil, a savings of nearly $63 billion dollars since 1973.

Current U.S. goals for the year 2000 are to reduce CO_2 emission by 108 million metric tons, N_2O by 2 million tons, and SO_2 by 10 million tons each year. According to NEI statistics, in 1993 alone, nuclear power plants helped utilities avoid the release of 133 million metric tons of CO_2, 2.2 million tons of N_2O, and 4.7 million tons of SO_2.

Scott Denman, executive director of SECC, questions the industry's claim that the use of nuclear power decreased the use of fossil fuels. He believes the decrease is due to a more efficient use of energy. Denman noted that during the 1978–1982 Arab oil embargo, utility companies reduced their use of oil by 60 percent, but "nuclear generated electricity remained virtually static while coal use increased 20 percent and electrical efficiency grew by more than 200 percent."[14]

The NEI solution for reducing the release of future emissions is to build more nuclear plants. Because nuclear power only generates electricity, however, SECC argues that it is not a realistic response to the environmental and supply problems of burning fossil fuels. Energy efficiency and renewable energy technology are the solution, according to SECC. Denman cites a Department of Energy (DOE) report showing that efficiency is already saving the equivalent of 14 million barrels of oil per day—over 5 billion barrels per year.[15] That represents a savings of $130 billion per year since 1973—more than 40 times what nuclear power has saved during the same 20 years.

If reduced pollution and dependence on foreign oil are our goals, SECC argues, then it is far better to require the auto industry to increase gasoline mileage and to install new energy efficient lightbulbs than to build dozens of new nuclear plants. If the gasoline efficiency of most cars increased to 30 miles per gallon,

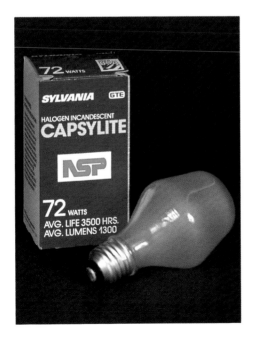

An energy-efficient lightbulb

the United States would virtually eliminate the need for oil imports.[16] According to SECC, energy-efficient lighting could prevent the emission of 232 million tons of CO_2.

Most nuclear energy opponents believe the choice of nuclear over fossil fuels merely represents trading one kind of pollution for an equally unacceptable kind—radiation.

WHAT DO WE KNOW ABOUT RADIATION?

Radiation is one of the chief forms of energy. Sunlight and minerals in the earth are natural sources of radiation. Artificially produced radiation is widely used in communications, medicine, industry, and research. Most scientists believe the amount of radiation to which we are exposed from sources created by humans carries little risk. Larger doses of radiation, however, cause biological and chemical changes in living tissue. Illness and even death can result if a person receives too much radiation.

The radiation associated with nuclear power plants consists of high-energy particles and rays produced by radioactive atoms. Radioactive atoms emit radiation in the form of alpha particles, beta particles, and gamma rays. They are called ionizing radiation because they can knock electrons off atoms or molecules, thus producing ions. Ionizing radiation can damage chromosomes in cells.

Alpha particles contain two neutrons and two protons. Because of their large size, alpha particles quickly interact with whatever they hit. A piece of tissue paper can stop them.

A beta particle is a high-speed electron. A thin sheet of aluminum can stop beta particles.

When the nucleus of a radioactive atom gives off an alpha or beta particle, the arrangement of protons and neutrons changes. This rearrangement produces gamma rays. Gamma rays are similar to X rays and can easily penetrate the human body. It can take more than three feet of concrete to stop gamma rays.

These types of radiation are emitted when a radioactive element, such as uranium, undergoes decay. The time it takes for half of a given amount of a radioactive substance to become stable is called the half-life. The half-life of U-235 is 700 million years. In other words, if we had a one-pound block of U-235, 700 million years later, half a pound would still be radioactive. After another 700 million years, half of that half pound would be radioactive, and so on. Each of the decay steps emits different types of radiation. Different radioactive elements have different decay chains and different half-lives. Once the process of fission is under way, over 200 different radioactive substances (or fission products) are created in the core of a reactor.[17] These fission products include plutonium 239 (half-life: 24,000 years), strontium 90 (half-life: 38 years), and iodine 131 (half-life: 8.1 days).

Most of these products stay inside the fuel rods, but some of them seep through tiny cracks in the rods and get into the coolant. They are trapped in filters, or in the case of gases, are released into the air—as was the case at Pilgrim. The radioactive products trapped in the rods and filters are called nuclear waste. The ones that escape are the source of the routine emissions that

Many people voluntarily expose themselves to radiation in the form of sunlight.

nuclear power plants disperse into the air and water. In general, scientists agree that multiplying a fission product's (an element's) half-life by 20 gives a reasonable estimate of when it will no longer be dangerous. Plutonium 239, for example, takes about 500,000 years to decay to safe levels.[18]

When ionizing radiation passes through a human cell, one of three things can happen. First, and most likely, nothing will happen. Think of a single ray of radiation as a sunbeam and the distance between molecules as the distance between planets in our solar system. The chance that any one sunbeam will strike the earth is slim indeed. So is the chance that radiation will hit anything important as it passes through a living cell.

Second, damage can occur that kills cells. Cells die in our bodies all the time. They are replaced by new ones. But if we are exposed to a massive amount of radiation, so many cells may be killed that the body's organs or tissues break down. This results in radiation sickness like that experienced by people after the atomic bombing of Hiroshima and Nagasaki.

Third, damage that causes cells to grow abnormally can occur. This kind of damage can result in cancer or genetic mutation. Cells may lose the ability to regulate their reproduction. They may multiply rapidly, causing cancer. Cancer can take five to thirty years (or more) to become evident after the initial alteration. This delay is called the latency period. In reproductive cells in an ovary or testis, the damage can cause genetic mutation—permanent change—in offspring. The damage

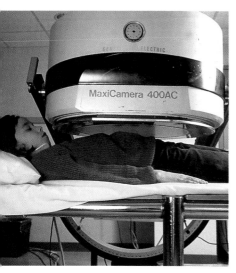 

Many everyday objects, such as x-ray machines, televisions, and smoke detectors, emit some radiation.

may not appear immediately. It may be carried through many generations.

Various types of ionizing radiation produce different degrees of biological damage. The relative biological effectiveness (RBE) of a particular radiation indicates the extent to which it damages cells compared to equal doses of other ionizing radiation.

The different types of ionizing radiation that can damage the human body are measured in rems. Another common unit of measure is the millirem, which equals 1/1000 of a rem. The millirem represents being struck by about 7 billion particles of radiation.[19] The average annual radiation dose received by a person in the United States is about 300 millirems. Much of this dose is due to natural radiation and medical and dental X rays. A small amount comes from nuclear weapons tests. The rest comes from color TV sets, smoke detectors, airport baggage X-ray machines, and nuclear power plants.

A person can receive a dose of up to 25 rems of radiation without showing any immediate effects. A dose of 100 rems may cause radiation sickness. Exposure to 300-800 rems causes severe radiation sickness. Doses above 800 rems are fatal.

Those who believe nuclear energy poses little threat to human health rely heavily on average dose charts like this. The charts indicate the tiny contribution the entire nuclear fuel cycle makes to overall exposure. But the Pilgrim plant situation leads others to argue that the risk to people near nuclear facilities is too great and our exposure standards are not protecting human health.

DO THE FEDERAL EXPOSURE STANDARDS REPRESENT AN ACCEPTABLE RISK?

Fear of radiation is a powerful emotional force that colors many people's response to nuclear power issues. Perhaps that fear comes from the association with images of death and destruction caused by nuclear bombs. Radiation is invisible to our senses: we cannot feel it, taste it, see it, smell it, or hear it. Recent reports about government secrecy and deliberate exposure of the public to fallout during early tests undermine public confidence in government assurances that radiation releases present "no danger to the public."

Whatever the origins of this fear, advocates of nuclear energy try to dispel it. Information pamphlets supplied by the NEI and used widely in the information centers at nuclear power plants stress that we live in a "sea of radiation" and that nuclear power represents only a tiny fraction of that yearly exposure. The NEI claims that "...the radiation from natural and man-made sources represents little or no risk."

In contrast, SECC notes that the National Research Council's Committee on the Biological Effects of Ionizing Radiation supports the linear theory.[20] According to the linear theory, any exposure to radiation carries some risk of causing harmful effects. This risk gets less and less as the dose grows smaller, but it never becomes zero. The theory also predicts that a small dose of radiation spread out over a large population can have serious consequences for human health. The Environmental Protection Agency studied the effects of radon gas, a natural source of radioactivity that seeps into buildings from the ground. Applying the linear theory, the EPA estimates that radon causes between 7,000 and 30,000 deaths each year.[21]

Some scientists insist that the linear theory itself may be wrong, and a threshold exists beyond which no harm occurs or where harm diminishes dramatically. Other scientists believe the risk actually increases at low doses. Scientists on both sides of the debate draw on the same data. Their interpretations of the meaning differ. They may accuse each other of bias, faulty methods, or distortions, but they rarely accuse each other of lying.

A consensus about the health risks associated with radiation exposure is established by scientific organizations such as the International Commission on Radiological Protection (ICRP) and the National Council on Radiation Protection and Measurement. Periodically, these organizations review the ongoing health studies of the 284,000 survivors of the atomic bombings of Hiroshima and Nagasaki, as well as many smaller studies of people and animals exposed to radiation. Regulators around the world then translate the standards into exposure limits. The Nuclear Regulatory Commission's standards are: 100 millirems per year average exposure for the entire population; up to 5,000 millirems for workers in nuclear power plants. Exposure at the fence around a nuclear power plant cannot exceed 5 millirems.

The yearly exposure from nuclear power is less than one millirem for the average American. The currently accepted consensus is that exposure to 1 millirem of radiation may increase your risk of getting a fatal cancer by one chance in two million.[22] It is also correct to say that if two million people are exposed to 1 millirem each, then one person will get cancer.

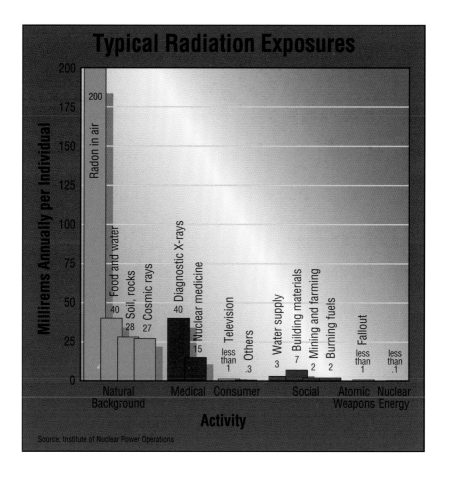

Scientist Bernard Cohen has studied everyday risks in society. He concluded that the individual risk from exposure to one millirem of radiation is comparable to taking three puffs on a cigarette, driving an extra three miles in an automobile, or an overweight person eating 10 extra calories.[23] The majority of people who support the use of nuclear power believe these risks are so small compared to other everyday risks that no informed person would give them a second thought.

But no matter how small the risk might appear, would you permit someone to force you to take these puffs on a cigarette? In the view of Dr. John Gofman, this is what the government allows the nuclear industry to do. In 1969, Gofman's research revealed that if the dose levels to the public ever reached the maximum permissible levels, approximately 32,000 Americans would die each year from the routine radioactive releases from nuclear power plants.[24] Since that time, Gofman has been a major dissenting voice in the field of radiation risk estimates. He believes official risk estimates are too low. His latest research suggests cancer risks from radiation may be 30 times higher than those estimated in the National Research Council BEIR V report published in 1990.[25] Gofman wrote in his book *Poisoned Power* that "the so-called permissible dose of radiation, for nuclear workers or for the public at large, represents only a legalized permit for the nuclear industry to commit random, premeditated murders on the American population."[26]

Gofman is rejected by nuclear power proponents as an alarmist. NEI asserts that "there is no evidence that humans are harmed by exposure to radiation below 10,000 millirems."[27] Thus, they believe the current NRC standards are extremely protective of human health and actual exposure levels are a hundred times less than permissible limits.

Credentialed experts, scientists, and researchers sharply disagree with one another. Decision makers in elected government, regulatory agencies, pronuclear and antinuclear groups must eventually choose whom to trust as they make their judgments concerning the

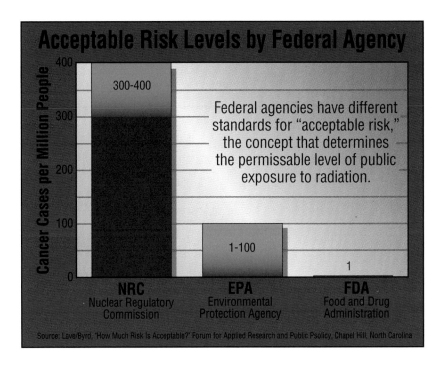

Acceptable Risk Levels by Federal Agency

Federal agencies have different standards for "acceptable risk," the concept that determines the permissable level of public exposure to radiation.

Cancer Cases per Million People

NRC — Nuclear Regulatory Commission: 300-400

EPA — Environmental Protection Agency: 1-100

FDA — Food and Drug Administration: 1

Source: Lave/Byrd, "How Much Risk Is Acceptable?" Forum for Applied Research and Public Psolicy, Chapel Hill, North Carolina

risks associated with very low levels of radiation exposure. You too will have to make a decision about the
risks of radiation. But whatever conclusion you reach,
it is a fact that only a small percentage of the total radiation humans have created since the dawn of the
atomic age has escaped into the environment so far.

Miles of pipes carry water throughout a nuclear power plant. The water transfers heat from the reactor to make steam, which generates electricity.

NUCLEAR SAFETY

The goal of nuclear safety is straightforward—to prevent harmful doses of radiation from reaching the public both when a nuclear power plant is operating properly and when an accident or a natural disaster strikes. While proponents and opponents of nuclear power argue about the effects of small, routine releases of radiation from a nuclear power plant, they do agree that the potential exists for serious accidents. They do not agree, however, about the likelihood of such accidents occurring.

The NEI says that a combination of safe reactor designs, excellent operating and maintenance practices, and vigorous oversight by the NRC assure a high level of safety. This combination of protections is called "safety-in-depth." Nuclear reactors are designed with sensors to watch temperature, pressure, water level, and other indicators important to safety. These sensors are linked to control systems that immediately adjust or shut down the reactor at the first sign of trouble. People who design reactors assume that sometimes equipment will fail and workers will make mistakes. For this reason, each control system has one or more back-ups. For instance, the Vermont Yankee plant has two emergency core cooling systems that are completely independent of each other. Either one can cool the reactor in an emergency. Finally, the entire reactor is enclosed in a massive, steel-lined, concrete building called the containment. The walls of the containment are four feet thick. The building is designed to withstand the heat and pressure resulting from an accident. It also prevents any radiation from leaking into the environment.

Because a nuclear plant is an extremely complex system of piping, wiring, and machinery, safety also depends on the people who operate and maintain it. A safety system that is broken or an operator who doesn't follow procedures creates unsafe conditions. Workers

Inspectors from the Nuclear Regulatory Commission carefully monitor equipment and review plant operations.

at a nuclear reactor receive rigorous training, pay attention to details, and are extremely disciplined. Nuclear plants receive careful maintenance. Old equipment is upgraded or replaced. Because of this constant attention, the industry claims that nuclear power plants never grow old.

The federal Nuclear Regulatory Commission oversees every nuclear plant in the United States. The NRC has safety inspectors at every plant and conducts regular reviews of operations. The commission also sets safety policy based on ongoing research and industry experience. The Nuclear Energy Institute contends that no other industry is so carefully and strictly regulated. The "safety-in-depth" approach has led to a level of safety and excellence unmet in other industries. Proponents ask what other major U.S. technology can point to its worst disaster—the accident at Three Mile Island—and say no one was injured.[1]

Critics of nuclear power say "safety-in-depth" sounds good in theory, but when that philosophy is applied by fallible human beings, the results are not always reassuring. According to SECC, half the containments surrounding nuclear reactors in the United States are not strong enough to keep radiation from escaping into the environment during a meltdown. The General Electric Company designed both the Pilgrim and Vermont Yankee containments. To save money on construction, the company built smaller containments that relied on cooling ponds of water or ice to absorb the heat and pressure of an accident. Engineers miscalculated, however, and NRC studies have revealed that such containments are "virtually certain to fail during a meltdown," and may lead to circumstances as severe as those in Chernobyl.[2] Although the NRC has known about these flaws, it has allowed the reactors to continue operating.

SECC also points out that strict inspections and maintenance regulations aren't necessarily effective ways to ensure safety either. The NRC requires inspection of reactor pipes, based on the theory that a weakened pipe will leak before it breaks. When an operator or inspector finds a leak, the reactor can be shut down before a serious accident happens. But on December 9, 1986, a water pipe broke at the Surry plant near Williamsburg, Virginia. More than 30,000 gallons of hot water poured out, flashed to steam, and scalded eight workers, four of whom died.[3] The pipe was one-half inch thick when installed, but it had thinned to one-sixteenth of an inch—thinner than a credit card. The pipe did not leak before it broke.[4]

SECC says that aging equipment can lead to a gradual reduction in the effectiveness of safety systems. For example, the protective coatings on wires can disintegrate in the 20 years since they were installed. This could lead to a short circuit and the failure of power to reach a critical safety system during an accident.

The NRC routinely identifies safety issues, such as aging wires or corroding pipes. If they apply to all reactors, they are labeled "generic issues." Once an issue is labeled generic, reactors can operate even if they have that particular safety problem. SECC says the gap between "safety-in-depth" and actual conditions at nuclear power plants has led to more than 30,000 mishaps between 1979 and 1989.[5] Three thousand of them occurred the same year Chernobyl exploded.[6]

Defenders of nuclear technology reject SECC's evidence as alarmist scare tactics. They assert that the account of mishaps is actually proof that nuclear safety is working. In none of these cases was the public in danger. NEI notes that the majority of the mishaps involved trivial incidents such as a fuse blowing, a seal leaking, or a valve sticking during a test. The careful reporting of such events reflects the industry's serious attitude toward safety. Even these trivial occurrences are analyzed for their safety implications.

Advocates of nuclear power accuse critics of demanding absolute safety when "nothing in nature or society is totally without risk," and the industry sees its task as "keeping that risk at an acceptable level."[7] But what is acceptable, and who should decide?

YANKEE ROWE: 30 YEARS YOUNG OR WORN OUT?

When the Yankee Atomic Electric Company in Rowe, Massachusetts, began operation in 1960, the 185-megawatt reactor helped pioneer the era of utility-owned nuclear power. In 1990, Yankee Rowe celebrated 30 years of operation. It was the oldest commercial nuclear plant in the United States, and it boasted one of the best operating records in the industry. In 1991 Yankee Rowe was again preparing to break new ground by applying for the first license renewal. Nuclear plants are licensed to operate for 40 years, and the Atomic Energy Act provides for a 20-year renewal. Over 30 nuclear plants will have to renew their licenses before the year 2010. License renewal is a lengthy process that involves a review of the entire nuclear

The Yankee Atomic Electric Company in Rowe,
Massachusetts, was shut down in 1991 after a regulatory
battle about whether the plant was still safe after 30 years
of operation.

power plant. Both the NRC and Yankee Rowe wanted to get the process underway early so they could thoroughly explore all the issues that might arise.

But on June 5, 1991, two citizen "watchdog" groups petitioned the NRC to shut down the Rowe reactor immediately. One of the groups, the Union of Concerned Scientists based in Cambridge, Massachusetts, is composed of scientists and citizens concerned with issues of nuclear safety, nuclear weapons, and energy policy. The other group, the New England Coalition on Nuclear Pollution (NECNP) is a regional citizens' group based in Brattleboro, Vermont. It has opposed the use of nuclear power since 1971. Both groups said NRC documents revealed that 30 years of radiation exposure from the core had so weakened the steel in the reactor vessel—a condition known as embrittlement—that it could burst open in an accident. The reactor vessel contains the radioactive core of nuclear fuel. It has no backup safety system. If it burst, emergency cooling systems could not keep water around the core. The water would flow out of the broken reactor, leading to a meltdown of the fuel and the release of radiation into the environment. Embrittlement of the steel sets up a condition in which the very system designed to save a reactor—the emergency core cooling system—could destroy it. In the event of a pipe break, the cold water might cause the hot, weak steel to shatter like a hot glass dropped into ice water.

Robert Pollard of the Union of Concerned Scientists contended the NRC knew the reactor was dangerously brittle, but it was allowing Rowe to operate with a safety risk that violated its own regulations. Pollard is a nuclear safety engineer and former NRC inspector, widely respected as the NRC's sharpest critic.

Yankee Rowe engineers maintained that the reactor was operating with a wide margin of safety. They calculated the risk of a rupture at 5 chances in one billion.[8] In comparison, the odds that a person will be killed in some kind of accident are 1.3 chances in a million.[9]

NRC's own engineers evaluated the available information differently. They estimated the risk of an accident at Rowe to be between 1 in 10,000 and 1 in 100,000. NRC officials admitted the estimate fell short of the NRC's safety goal of 5 in 1 million, but they rejected the citizens' petition. At a public meeting, NRC officials told over 500 citizens that operating the reactor "did not pose an undue risk."[10]

The wide discrepancy between Yankee Rowe's figures and NRC estimates was at the heart of the controversy. Yankee Rowe should have been able to prove that the reactor vessel was safe. Metal samples exactly like the steel of the reactor had been placed inside it when the plant started operation in 1960. These samples were to be removed periodically and tested, but they were lost early in the reactor's operation. This forced Yankee Rowe and the NRC to rely on computer models and statistics—sometimes using as few as three pieces of real data. The only way to be absolutely certain of the reactor's safety was to have a special robot scoop a metal sample out of the inside and test it. But the robot wouldn't be ready for a year. If the NRC enforced its regulations, the power plant would have to sit idle for all that time.

Many people felt the NRC's judgment about safety had been influenced by a desire to protect the utility's investment. They thought the safest thing to do was shut down the reactor until the robot was ready. One citizen said, "Nobody asked us, do we want these risks?"[11]

Two events heightened public concern and prevented the issue from ending with the NRC's rejection of the petition. Lightning struck the plant on June 15. The damage disabled the generator and forced an emergency shutdown of the reactor. Because emergency notification systems failed to work properly, some area towns did not hear of trouble at the plant for nearly an hour. Plant officials had told people they would be notified of an emergency at the plant within

15 minutes after it occurred. Residents began to question the evacuation plans, as they thought about actually having to use them. They were alarmed to discover that there weren't enough school buses to evacuate children if an accident happened on a school day.[12]

Meanwhile, a dissenting voice emerged from the NRC itself. N. Pryor Randall, a metallurgist with the NRC, said he could not agree that operating Rowe was "safe or justified."[13] As the uncertainties mounted, tensions increased in area communities. Resolutions for and against Rowe were put to vote. People who worried that the plant posed a danger had neighbors who worked at the reactor. Communities were being torn apart. Some believed Yankee Rowe was a "good neighbor," while others thought "...the vision of a nuclear accident looms in our minds."[14]

Federal and state representatives responded to public concern and pressured the NRC to reconsider its decision. One NRC worker said the agency had never before received so many calls on an issue.[15] The NRC reconsidered the petition three times, and on October 1, 1991, decided the reactor should be shut down until tests proved it was safe. The utilities that owned Rowe decided it would be too costly to get the answers the NRC wanted, and they closed the reactor—forever.

SAFE ENOUGH?

There might never have been an event at Rowe to challenge the reactor vessel. The reactor might have been safe enough. Nuclear safety involves the anticipation and prevention of dangerous conditions before accidents happen—so it will always involve human judgments subject to criticism. The Rowe controversy exemplifies the major elements of the safety debate: the safety of the reactor, public confidence in regulators, and personal judgments about acceptable risk.

Dr. Anthony V. Nero, a nuclear physicist, wrote a popular textbook called *A Guidebook to Nuclear Reactors*. He was asked to contribute his views on nuclear reactor safety to the book *Both Sides*, which is a collection of essays by pronuclear and antinuclear experts. Nero wrote: "The only practical requirement for nuclear power or any other energy technology is to diminish the chances of a big accident. If careful design and operation make the accident probability tiny enough, then nuclear reactors are adequately safe."[16]

Nero based his confidence on the results of the first and most famous attempt to put numbers on nuclear accident risks—a study known as WASH-1400. Dr. Norman Rasmussen and a team of experts made the $3 million study at the request of the Atomic Energy Commission. Using probabilistic techniques developed to calculate rocket failures, WASH-1400 identified the ways nuclear safety systems could fail and how likely they were to do so. It then tried to account for human error. The results were reported as the chance of a severe accident for each year that a reactor operated, known as a reactor-year. If 1 reactor operates for 10 years, then 10 reactor-years have accumulated. Similarly, if 10 reactors operate for 1 year, then 10 reactor-years have accumulated.

WASH-1400 concluded that the chance of an accident that could release enough radiation to cause 3,000 deaths was one in one-half million reactor-years. Put another way, if 100 reactors are operating in the United States (about the number in operation at the time of the Three Mile Island accident), then we might expect one serious accident every 5,000 years.

Nuclear power supporters often say the public fails to consider risks as scientists do. People become upset by the large number of possible deaths while failing to consider how unlikely they are to occur. The "average" risk concept means that if a nuclear accident might kill 1,000 people once every million reactor-years, then the risk to society is one death every 10 years.

This risk is far less than the zero to eight deaths per year caused by air pollution from each coal plant. Nero

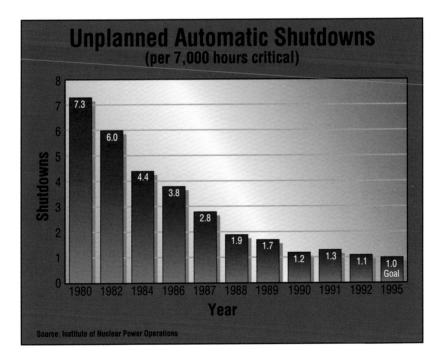

wrote, "Though the trauma of such an accident would be great, the net risk to society is small compared with other risks we accept routinely." He noted that airline accidents can also kill hundreds of people, and dam failures could kill thousands or tens of thousands of people. Nero argued that when the very small odds of an accident are combined with the seemingly severe consequences, then "...the *average* risk from nuclear power is no more, and probably much less, than the risk from burning coal."[17] He concluded that nuclear power is "indeed, safe enough."[18]

Jan Beyea, a staff physicist with the National Audubon Society, studied the dispersion of radioactivity from a simulated meltdown. He, too, was asked to contribute to *Both Sides.* Beyea was critical of the faith scientists such as Nero place in risk assessment. Beyea suggested the nuclear industry was so new that making predictions was like someone at the beginning of the automobile age in 1905 trying to guess how many traffic deaths would occur in 1990. Anyone who had suggested 50,000 deaths would have been called an alarmist, the position nuclear critics found themselves in at the beginning of the nuclear age.

A nuclear plant is a complex collection of thousands of systems and parts built by dozens of different companies, designed by hundreds of engineers, and operated by about 80 different utilities. Beyea questioned whether engineers could "...visualize all possible behavior of their systems, all possible defects in the design and construction, all possible operator malfunctions, all possible activities of madmen and terrorists."[19]

He said the best way to see how overly optimistic

studies such as WASH-1400 can be is to look at past accidents. For example, an electrician testing for air leaks with a candle caused a 1975 fire at Browns Ferry that put two reactors dangerously out of control for 16 hours. The subsequent fire destroyed thousands of control cables for both reactors. Fires, Beyea pointed out, were not even considered in the WASH-1400 study. Browns Ferry demonstrates that safety engineers cannot protect against what they have not imagined.

The Three Mile Island accident might not have occurred for several decades. Only 1500 reactor-years had accumulated worldwide when the partial meltdown occurred in 1979. TMI itself had only been in commercial operation for three months when it was destroyed. Beyea concluded designers had missed their safety risk goal by a factor of 500 and that "...at some point we have to give the nuclear designers a failing grade."[20]

Beyea believed that "many rational people tolerate the uncertain risk of a nuclear accident in preference to the actual number of deaths from coal facilities which occur yearly."[21] But unlike Nero, Beyea also believed that if just one major reactor accident occurred, "then the total damage to human health and the environment from this one event would make nuclear power a questionable bargain."[22] Beyea and Nero were writing in 1982. Four years, and only 1900 reactor-years later,[23] Chernobyl exploded.

But odds can be tricky. If an event is predicted to happen once every 100 years, that means twice every 200 years. Both events could happen on the same day, however, and still agree with the odds.

LESSONS LEARNED

Nuclear reactors follow the laws of physics, but the human beings who build, regulate, and operate them are subject to different, less predictable forces. In 1969, physicist Edward Teller, known as the father of the hydrogen bomb, said about nuclear power: "So far we have been extremely lucky.... But with the spread of industrialization, with the greater number of simians monkeying around with things they do not completely understand, sooner or later a fool will prove greater than the proof even in a foolproof system."[24] Teller wasn't against nuclear power. He was just stating the obvious, and he was right.

In Middletown, Pennsylvania, at 4:37 A.M. on March 28, 1979, the reactor at Three Mile Island Unit 2 was running at full power. A mistake by maintenance personnel caused the electric generator to shut down. Since the heat was no longer being carried away from

Edward Teller warned that human error would increasingly become a factor in nuclear power plant accidents.

the reactor, the pressure rose inside the pipes filled with coolant. Within four-hundredths of a second, a pressure relief valve opened to protect the pipes from the increase in pressure—just as it was designed to do. The valve should have closed in 13 seconds. It didn't, but a control room light falsely told the reactor operators that it had. Since the indicator light did not reflect the real state of the valve, operators did not know that 220 gallons of radioactive coolant were spilling out of the stuck valve every minute.

One-tenth of a second after the generator shut down, the reactor safety systems responded to the emergency as they were designed to do. Control rods were automatically inserted, stopping the fission reaction. Two minutes later the emergency core cooling system switched on. The automatic start-up of the ECCS indicated that the water level in the reactor had become dangerously low. If the water level drops below the top of the core, a meltdown will begin. But the operators couldn't directly measure the water level. They thought the reactor was full of water, so they overrode the automatic safety system. Once the ECCS was shut off, the worst nuclear accident in the United States became inevitable. Two hours later, a shift change brought a new operator onto the scene. He recognized that the relief valve was stuck open and closed it—only an hour before total disaster would have occurred.[25]

Even when the cause of the accident was discovered, several days passed before the reactor was stabilized. Radiation—including radioactive iodine—escaped almost continuously because of leaks. Federal and state officials didn't decide to order an evacuation

The four cooling towers at the Three Mile Island nuclear power plant dominate the horizon. The plant was shut down after a serious accident occurred in 1979.

until three days later, but by then more than 100,000 people had already fled. The reactor had been destroyed. Fifty percent of the core had melted, and the containment was filled with hundreds of thousands of gallons of radioactive water. TMI Unit 2 could never be used again. Cleanup took 10 years and cost $1 billion. The surrounding communities were left with lingering fears of illness.

Ruth Hoover, a dairy farmer near TMI, related an experience common to many during the accident: "They say you can never taste radiation. Okay, that's true, but you sure can taste the iodine. We had the bitter taste in our mouths."[26]

TMI demonstrated that the antinuclear voices had not been crying wolf after all: nuclear accidents really can happen. In the minds of nuclear power proponents, TMI also demonstrated that "safety-in-depth" works. Despite gross abuses, a reactor can suffer a severe accident without killing or injuring anyone. Critics note that TMI was forced to the last line of defense—the containment—and even that leaked, exposing citizens to undetermined levels of radiation.

However, NEI says the average dose of radiation was less than 1.5 millirems to people within 50 miles of the plant.[27] NEI maintains that no one was injured and that "...at least a dozen epidemiological studies between 1981 and 1991 have found no discernible direct health effects to the population in the vicinity of the plant." Several of these studies were conducted by the Pennsylvania Department of Health, and no effect other than stress has been demonstrated.[28]

The reactor, inside the containment building, looks blue because it is submerged in water.

Eight-year-old David was a victim of stress. According to an account in the book *People of Three Mile
Island,* David had been evacuated during the accident.
For a long time after returning home, he struggled with
constant anxiety, clinging to his mother like a three-
year-old. He had to call home from school twice a day
to ask "Is everything all right? Is everybody still alive?"
Then, one day he came back from visiting a friend to
find his house empty. The family had only gone shop-
ping, but his aunt said David "...just cried and cried
and sobbed and sobbed his heart out."[29]

Though the issue of health effects remains highly
controversial, few would deny that TMI marked a time
when "...the nuclear power industry lost its inno-
cence...and [its] credibility suffered a partial meltdown."[30] Beyea wrote that until TMI, the industry and
government regulators "treated the possibility of cata-
strophic accidents as a public relations problem, not a
safety problem."[31] Nor did utilities realize the true
stakes: the failure of a fifteen cent part could ruin a billion dollar investment.[32]

A presidential commission headed by Dartmouth
College president John Kemeny investigated the accident and concluded that the primary causes were
human indecisiveness and errors in the control room.
These factors were compounded by the utility's inade-
quate attention to maintenance and training, and flaws
in the NRC's and industry's attitude toward safety. The
Kemeny report concluded: "No amount of technical
fixes can cure this underlying problem."[33]

Regulators and industry officials argue that TMI was
a wake-up call that forced the industry to get in shape.

NEI says that "the industry not only complied with the recommendations of the Kemeny Commission but also transformed the way it manages and operates plants."[34] They note that Kemeny observed on the tenth anniversary of the accident that "the [industry's] improvements over the past decade have been impressive and are very reassuring." As a result, NEI says, nuclear plants operate "...more safely, more efficiently, and more reliably from year to year."[35]

But in spite of all the improvements, the nuclear industry did not emerge undamaged. The future growth of nuclear power was stalled. Not one new plant has been ordered since 1974, and over 100 reactors already planned were canceled in the wake of the Three Mile Island accident.[36]

Opponents also doubt the lessons of TMI or Chernobyl have been learned. The Rowe controversy, where the NRC had to be pressured to enforce its safety regulations, indicates that safety continues to be a last concern. According to a SECC article, as late as 1989 (10 years after the TMI accident) six out of eight plants with designs similar to TMI still had not complied with post-TMI safety modifications. Furthermore, only 12 out of 112 reactors had installed NRC-approved Safety Parameter Display Systems that could help avoid the kind of control room confusion that contributed to the TMI meltdown.[37]

Scott Denman writes, "Unbelievably, in the wake of the Chernobyl disaster, the NRC has proposed to roll back emergency planning...," despite their own reports stating that all types of U.S. containments can fail during a severe core meltdown.[38]

POLICY MAKERS AND CITIZENS

The work of the Kemeny Commission was applauded by the White House, the NRC, and the nuclear industry. Yet the image of the complete safety of nuclear power had been shattered, and protests sprang up across the nation with calls to bring a halt to its development. Over 100,000 people protested in Washington, D.C. Meanwhile, hearings were underway to allow TMI Unit 1 to reopen—and maybe even Unit 2. (No one at that time understood that the damage to Unit 2 was too great to ever permit its operation again.) Citizens who had already been through the terror of one accident tried to stop the restart of Unit 1.

At a town meeting to discuss a resolution against the restart of TMI, a man held up a child and someone asked, "How many watts is that kid worth? How many jobs is that kid worth? We're here because we feel endangered."[39]

Dr. Judith Johnsrud is a veteran activist who had successfully participated in efforts to stop more than eight nuclear projects in Pennsylvania, including the cancellation of seven reactors.[40] She said six local citizens' groups organized to intervene in the hearings. They fought for two years to prevent a restart at Three Mile Island. "But they overwhelmed us. You've got to realize the purpose of an NRC preceding of any kind is to *approve* the license. It is not to consider whether or not to approve."[41]

Regulators, however, were not as hasty as Johnsrud implies. The NRC placed a moratorium on new licenses and construction permits while it studied the TMI accident, and it developed new regulatory

*At a town meeting, citizens
protested the reopening of
TMI Unit 1. As a man held
up his child, someone
asked, "How many watts is
that kid worth?" Below,
Reginald Gotchy and John
Collins of the Nuclear
Regulatory Commission try
to reassure the group.*

measures to prevent future accidents. In 1985—six years after the accident—the NRC finally decided TMI Unit 1 was ready to restart.

Officials in Chernobyl reacted to their accident differently. Human error and design flaws were at the center of the 1986 Chernobyl accident. The containment building could not withstand the violent explosion, and deadly amounts of radiation escaped into the atmosphere. Did Soviet officials do enough to protect the people living within the area contaminated by radiation? In 1989, the Ukrainian minister of health, A. Romanyenko, stated that "no one is suffering today from illness due to radiation poisoning."[42]

According to investigative journalist Alla Yaroshinkskaya, tens of thousands of people who should

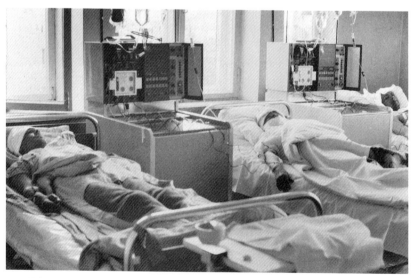

Victims of the Chernobyl nuclear disaster lie in a hospital. They are suffering from severe burns and internal damage from radiation.

have been evacuated to protect their health were not. Yaroshinkskaya reported in her book *Chernobyl: Forbidden Truth* that for years after the accident, people living in contaminated zones were not informed. They grew crops, drank milk, and allowed their children to play in the radiation-laced dirt. Instead of evacuating the population, officials increased levels of permissible radiation doses 10 to 50 times above what they had been before the accident.

Romanyenko was later forced to agree that "1.5 million children have received worrying doses of radiation."[43] Local doctors in the country of Belarus, where the majority of the fallout landed, report nearly one-third of all children are sick, and rates of bone, blood, and nervous diseases and tumors have nearly doubled. The people fear the extinction of their nation.[44]

NEI calls into question the significance of accounts like this by emphasizing reports from the International Red Cross, the World Health Organization, and the International Atomic Energy Agency that say "...people in the affected areas are living under such stress that they attribute virtually all illness...[to Chernobyl], even when radiation is plainly not the cause."[45] Some people suggest that people were sick before the accident, and it is the increased medical attention that is revealing illnesses not caused by radiation from Chernobyl. The NEI's literature states that 200 people were injured, and 30 died as a result of Chernobyl.

Disputes about the health effects from the accident will probably continue for generations. One irrefutable consequence, however, is the dislocation of thousands of people, the contamination of prime croplands, and billions of dollars in cleanup costs. It seems reasonable to wonder whether the United States would be any more capable of carrying out a massive relocation effort if a similar disaster were to occur here. Could farmers afford to destroy contaminated crops and pour radioactive milk down drains, or would we be forced, like the Soviets, to mix these in with the national food supply?

Antinuclear activists want to shut down reactors before we ever have to find out. Proponents, however, agree with Frederick Seitz, former head of the National Academy of Sciences (NAS). Seitz says, "Today, the chances of a meltdown that would pose a threat to U.S. public health are very slim."[46]

INFLUENCING PUBLIC PERCEPTIONS OF SAFETY

Many things clamor for your attention—like passing tests, getting into band, making the football team, or choosing a date. If you listed the 10 most important things on your mind, nuclear power probably would not be one of them. Your parents, too, have their priorities. Polls consistently show that energy is not among the top 10 issues on people's minds.[47]

In our increasingly complex world we must respond to more than immediately visible dangers. If human curiosity did not probe the unknown, we would never have perceived the rise in atmospheric carbon dioxide that is causing global warming and would never have discovered the ozone hole. Scientific progress and human survival often depend on new discoveries.

Ruth Hoover, the dairy farmer near TMI, said, "People outside the...50-mile radius, they don't understand. It's easy for them to forget. But if they would have put it [information about radiation] in the papers and educated the people, Three Mile Island or any nuclear reactor would never have been built in the whole United States."[48]

Antinuclear activists see their role as providing that education. Debbie Katz is president of the Citizens' Awareness Network, a group that emerged from the Rowe controversy. She described how knowledge leads people from apathy to activism. "Awareness changes everything. Once you begin to question how a community is affected by the standard operation of a reactor, you become aware of what's involved in the nuclear fuel cycle, and that changes everything."[49]

Supporters of nuclear power consider this the kind of thinking typical of hysterical activists who exaggerate the dangers of nuclear power—not to educate the

Groups such as the Safe Energy Communications Council (SECC) and the Nuclear Energy Institute (NEI) publish pamphlets promoting their views about nuclear power.

public, but to advance their antinuclear agenda. Supporters contend that knowing the truth about the risks of radiation and possible accidents would eliminate public fears, not increase them. In an attempt to reach the public, industry groups like NEI use full-page ads in major newspapers and national magazines, such as *National Geographic.* Images in these ads often include kids eating ice cream, a grandfather showing a leaf to his grandson, and kids fishing near nuclear plants. NEI also publishes a quarterly magazine called *Nuclear Industry,* maintains a speaker's bureau, and lobbies Congress. These information campaigns are backed by millions of dollars from utility companies and other members of NEI. In addition, every nuclear plant has a full-time public relations staff.

The nuclear industry clearly has more money for publicity than grassroots citizens' groups, but what about getting the kind of mass media attention that is critical to influencing public opinion? In the view of Bernard Cohen, activists have the advantage of a sympathetic media that gives the impression that radiation is far more dangerous than it is. Mass media sensationalizes accidents like TMI, which Cohen believes hurt no one. His study of major newspaper stories between 1974 and 1978 revealed only 120 items relating to traffic accidents, which killed 50,000 people each year, compared to 200 items on radiation accidents, which caused no deaths.[50]

In a *Nuclear Industry* article titled "Nuclear Images: Thanks for the Memories, Godzilla," author Hunt Williams speculates that the explosion of media interest in the 1970s resulted from a transference of anxiety about nuclear bombs to nuclear plants. The media's coverage is described as biased, critical, and alarmist, playing on public fascination with disaster and fear of all things nuclear.[51]

But according to Diana Sidebotham, hysterics and radicalism had little to do with the origins of the anti-nuclear movement. Sidebotham began to question nuclear power over 25 years ago during the licensing of the Vermont Yankee nuclear power plant. "On the contrary," she asserts, "it was scientists and engineers who spoke out from inside the nuclear industry. We heeded their warnings."[52]

Sidebotham is a trustee of the New England Coalition on Nuclear Pollution, an organization based near the Vermont Yankee plant. This citizens' organization is over 25 years old. With a handful of other groups, it can trace its origins back to the early days of nuclear power. Sidebotham recalls that the Atomic Energy Commission's licensing board for Vermont Yankee would not even discuss the problem of nuclear waste disposal: "They said that would be taken care of later. Well, now is later and we have no solution and a mountain of deadly materials."

But nuclear proponents think this kind of interference by citizens created the delays and cost increases that have crippled the nuclear industry. Frederick Seitz claims that "...opponents have so complicated its development that no nuclear power plants have been ordered or built here [in the United States] in 12 years." [53]

Activists point out that citizens have a democratic right to raise questions when they think there are safety problems that are being ignored. They say they did not build the nuclear power plants with faulty containments, create the safety and regulatory defects the TMI accident revealed, cause reactors to become embrittled, or design the equipment that is falling apart after only 20 years of use. According to Christopher Flavin of the Worldwatch Institute, these are the signs of an industry that grew too fast and paid too little attention to problems that would arise down the road.[54]

Citizens used a variety of techniques to get out their message and gain attention. During a major crisis, such as the Rowe controversy, some groups chose inflammatory, confrontational language and images. For example, after the TMI accident, "Better Active Today Than Radioactive Tomorrow" became a popular slogan.

Mass protests and nonviolent civil disobedience have occurred at reactor sites around the world. On June 25, 1978, more than 18,000 people protested at the Seabrook nuclear plant in New Hampshire.[55] Other groups chose complex technical and legal interventions in NRC hearings that required lawyers and large sums of money.

When there is no immediate crisis, nuclear power opponents use newsletters, house meetings, and fundraising events to persuade their neighbors that nuclear technology poses a threat to them. The Connecticut-based People's Action for Clean Energy (PACE) dramatizes what it sees as the danger of nuclear power with a poster that superimposes the Chernobyl impact zones on a map of New England. PACE includes this quote by former NRC commissioner James K. Asselstine on the poster: "The bottom line is that, given the present level of safety being achieved by the nuclear power plants in this country, we can expect to see a core meltdown accident within the next 20 years, and it is possible that such an accident could result in off-site releases... larger than...Chernobyl."

Activists argue that we are in this terrible situation because the NRC and the AEC never denied an operating license to a nuclear power plant once it was built. They accuse federal regulators of bending over backwards to serve the interests of the utilities at the expense of public safety.

Three Mile Island taught the importance of having evacuation plans for people living near nuclear power plants. New reactors cannot be licensed without an approved plan. The Seabrook Nuclear Power Station is located in Seabrook, New Hampshire, on the coast of the Atlantic Ocean and near a popular beach. Many local governments refused to participate in developing

Demonstrators protest the opening of the Seabrook Nuclear Power Station.

an evacuation plan because they thought it was impossible to evacuate thousands of tourists if an emergency occurred.

The NRC could not let Seabrook operate without a plan, so President Ronald Reagan ordered the Federal Emergency Management Agency (FEMA) to override the local governments and draw up a plan. On September 11, 1987, however, even FEMA concluded that thousands of people would be unable to escape radiation exposure.[56] According to SECC, the NRC demanded that FEMA reverse its decision. The FEMA official who would not approve the plan was replaced, and FEMA changed its position. In March 1990, the NRC granted Seabrook a license.[57]

Industry supporters don't believe the NRC favors nuclear power plant operators. They point out that the NRC has shut down several plants for safety reasons. For example, NRC inspectors closed the Peach Bottom reactor in Pennsylvania when they found reactor operators dozing while on duty in the control room. In the spring of 1996, the NRC ordered all four Connecticut reactors shut down because regulators thought the plants were no longer safe to operate. After the Yankee Rowe plant shut down, NEI spokesperson Scott Peters said, "It's always been my experience that the NRC has shut down plants that don't meet its qualifications."[58]

But activists remain skeptical. They note that some workers had been warning the NRC about problems at the Connecticut plants for several years. Then a highly damaging cover story appeared in the March 4, 1996, issue of *Time* magazine that implicated the agency in the safety problems. After the magazine article appeared, the NRC took action.

The safety question is likely to remain a subject of intense debate because, according to Christopher Flavin, scientists cannot answer the central philosophical question: "How safe is safe enough?" Only individual people working through the political process can resolve that issue.[59]

In the United States, an act of Congress would be required to stop the development of nuclear power. Congress would have to change the Atomic Energy Act, which makes it national policy to promote the use of nuclear power. No national movement is underway to persuade Congress to do this, but attempts have been made on the state level. According to the NEI, however, when Americans go to the ballot box, nuclear power comes out the winner. Of 24 voter initiatives against nuclear power between 1976–1992, only 3 succeeded. One plant shut down and two placed restrictions on new construction.[60] Despite the industry's success in winning voter support, many observers on both sides of the issue feel a halt to nuclear power may be on the horizon. It could be brought about by the equivalent of a "clogged toilet" as 50 years of nuclear waste accumulates without a safe place to dispose of it.

*Scientists conduct tests in this underground tunnel to learn
more about the Yucca Mountain site, which is proposed as a
possible national nuclear waste storage site.*

WASTE DISPOSAL

Ellen Jenkins's husband is alive today because nuclear technology has fulfilled one of its greatest promises— fighting cancer. Ellen said, "My husband had a brain tumor in 1989, and if it hadn't been for nuclear medicine, they wouldn't have found it."[1] The Jenkinses live in Barnwell County, South Carolina, the site of a nuclear waste disposal facility. Since 1971 nearly half of the nation's low-level nuclear waste has been disposed of there.[2] But Barnwell was scheduled to close in 1993. That might have left the nation without a single disposal site and Ellen worried what might happen to people like her husband if the activities that generate nuclear waste had to stop. "You know, a lot of the waste they bury is stuff like medical gloves and things like that. You can't tell someone who has a tumor, 'I'm sorry, we'd like to help you find where it is, but we can't because we can't dispose of the stuff from the hospital.'"[3]

According to an NEI fact sheet, nearly 10 million Americans are tested or treated with nuclear medicine each year. Eighty percent of all new drugs must be tested with radioactive materials before they can be approved for use. Aircraft manufacturers use radiation to check for flaws in jet engines. Farmers use radioactive materials to control pests and obtain better crop yields. Smoke detectors in your home and school rely on a tiny radioactive source to sense smoke and sound the alarm.

These are just a few of the ways in which over 20,000 licensed users of radioactive materials help make our lives safer, healthier, and more productive. These uses of radioactive materials also create nuclear waste that is classified by the Nuclear Regulatory Commission as "low level." A large amount of this low-level waste is clothing and equipment contaminated by contact with radioactive materials. It is usually safe to handle when packed in boxes or drums. Some of it, however, comes from nuclear plants and is so intensely radioactive that it would be deadly if it wasn't packed in heavy concrete and steel containers. The waste is sorted into three classes based on how long it remains dangerous: A = 100 years; B = 200–300 years; C = 500 years.

Until 1980, disposal of this waste drew little national public attention. Disposal took place at six landfills: Maxey Flats, Kentucky; West Valley, New York; Sheffield, Illinois; Richland, Washington; Beatty, Nevada; and Barnwell, South Carolina. Marvin Resnikoff is a physicist and waste consultant. He reported in his book *Living without Landfills* that the first three sites closed because of problems that included water seeping into the trenches, erosion of the dirt covers, and migration of radioactivity into ground water. NEI acknowledges there have been problems, but says, "...no

After being used for research, this 210-ton nuclear steam generator was decontaminated and painted to assure that no radioactive contaminants would be released into the environment. It is being transported for storage in Richland, Washington.

public health problem was ever identified with any of the sites,"[4] and the problems have been overcome with improved disposal technologies, such as special plastic liners and concrete vaults. Resnikoff, however, remains critical of the industry's claim that nuclear waste can be isolated from the environment, especially using the landfill method. He wrote, "Rather than being maintenance free [for the hazardous life of the waste]...these sites have ended up requiring active maintenance within 10 years of trench closure."[5]

Wastes had to be dug up, put in new containers, and reburied in new trenches. Radioactive water had to be collected to prevent contamination of nearby streams. The waste was supposed to have been safe to bury and forget; instead, it threatened the environment and increased worker exposures.

In 1980, the states containing the three remaining dumps—Washington, South Carolina, and Nevada—decided they did not want to be the dumping grounds for the entire nation. In response, Congress passed the Low-Level Radioactive Waste Policy Act (LLRWPA) of 1980. The new law required each state to take responsibility for the waste created within its borders. The states could develop their own disposal sites or form "compacts" with other states to share a site. LLRWPA represented a major nuclear waste disposal policy reversal by the federal government. What had once been a problem for private industry became the responsibility of state governments. Despite this law, by 1985, no state had formed compacts or chosen disposal sites.[6] Congress amended LLRWPA in 1985 to force the states to act by creating deadlines and penalties. The law also permitted the three existing disposal sites to close.

While citizens in other states organized to keep dumps out of their states, the people of Barnwell organized to keep their dump open. According to an article in *Nuclear Industry* magazine, town and county officials formed the Save Chem-Nuclear committee. Chem-Nuclear is the company that operates Barnwell. When the state of South Carolina held a hearing on whether to keep Barnwell open, 80 residents testified. According to the article, "No one from Barnwell County had a bad word to say about the low-level waste disposal site."[7] Those testifying considered the facility a good neighbor, providing local jobs and taxes to the town.

"They live right here," said school superintendent Carolyn Williams. "They have children here. They buy groceries. . . . they're out there, and they believe in what they're doing and know the safety element is there. That counts for a lot."[8]

Members of the committee traveled to the state capital and tried to have someone present at all the hearings that would decide Barnwell's fate. The citizens' efforts, in part, persuaded the legislature to keep Barnwell open. Their victory meant the county would keep $1.6 million in annual fees, gain an extra $1 million each year from surcharges, and not lose, in the words of Mayor Creech Sanders, "a good corporate neighbor."[9]

BENEFITS WITHOUT COSTS?

Faced with the LLRWPA amendments, most states formed compacts, while a few decided to find their

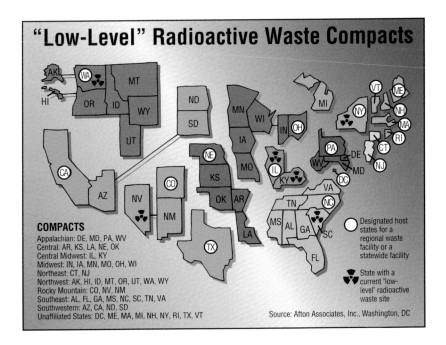

"Low-Level" Radioactive Waste Compacts

COMPACTS
Appalachian: DE, MD, PA, WV
Central: AR, KS, LA, NE, OK
Central Midwest: IL, KY
Midwest: IN, IA, MN, MO, OH, WI
Northeast: CT, NJ
Northwest: AK, HI, ID, MT, OR, UT, WA, WY
Rocky Mountain: CO, NV, NM
Southeast: AL, FL, GA, MS, NC, SC, TN, VA
Southwestern: AZ, CA, ND, SD
Unaffiliated States: DC, ME, MA, MI, NH, NY, RI, TX, VT

○ Designated host states for a regional waste facility or a statewide facility

☢ State with a current "low-level" radioactive waste site

Source: Afton Associates, Inc., Washington, DC

own solution. Almost all states passed siting laws and got the search process underway, although some, like New Hampshire, dragged their feet.

Submitting a license comes at the end of a process that begins with a search throughout the entire state or region for sites to locate the waste dump. A few sites are selected. Geologists, biologists, and engineers then study them carefully to make sure the water patterns and soil composition are favorable. They choose one of several disposal technologies—landfills, concrete vaults, or buried structures—and the license is submitted along with hundreds of pages of supporting documents. Either the state, the compact, or the NRC reviews the license. By 1996, sixteen years after the original law was passed, not one new low-level waste disposal facility was in operation. Four licenses had been submitted. One was granted in Ward Valley, California, and one was rejected in Illinois. Two were still being reviewed when this book was written.[10]

The NEI says failure to provide adequate disposal will bring a heavy price. We'll lose the advantages gained from medical and industrial uses of nuclear technology and jeopardize our ability to meet electricity needs. Dr. Conrad Nagle, president of the American College of Nuclear Physicians, has treated many patients with nuclear medicine. In an advertisement for the nuclear industry, he said "...I doubt they [his patients] have ever thought about the radioactive waste these medical procedures created. How typical of us all. We're glad to accept the benefits of radioactive materials—but not the responsibility for waste disposal.... Time is running out. For all of us."[11]

Resnikoff argued, however, that the nuclear industry was using the medical community to "run interference." The nuclear industry, according to Resnikoff, creates the impression that people will start dying without new disposal facilities when most medical waste decays in hours or days and it can be stored at hospitals until it is safe to throw away. Resnikoff pointed out that just one 5-megawatt reactor produces half of the radioactive materials used by the medical community. By contrast, a typical nuclear plant creates 3,500 megawatts total power. As a consequence, the 110 nuclear plants produce 99 percent of the total radioactivity in low-level waste. The rest, says Resnikoff, "...is only the pimple on the elephant."[12]

DIFFERENT VIEWS OF THE PROBLEM

Some antinuclear groups do believe in "clogging the toilet" in the hope that support for nuclear power will diminish. But activists did not create the issue that communities all over the United States are facing. The NEI says polls show people want waste sent to centralized facilities, but when the issue comes close to home, a different reaction occurs.[13] Are Americans irresponsible? Do they believe that the production of nuclear waste is okay as long as it's dumped in someone else's backyard? Is the effort to find dump sites being blocked by "small, politically savvy antinuclear groups," as Dr. Nagle claims?[14] Or is something more complicated going on?

Both sides agree in principle that isolation from the environment is the proper way to treat the dangerous radioactive waste. Water must not come into contact

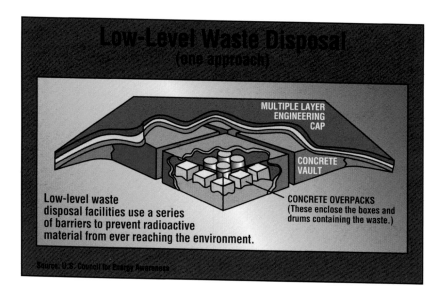

Low-Level Waste Disposal
(one approach)

MULTIPLE LAYER
ENGINEERING
CAP

CONCRETE
VAULT

Low-level waste
disposal facilities use a series
of barriers to prevent radioactive
material from ever reaching the environment.

CONCRETE OVERPACKS
(These enclose the boxes and
drums containing the waste.)

Source: U.S. Council for Energy Awareness

with the waste. People must be kept from intruding into a facility. The NEI emphasizes the low level of hazard and focuses attention on medical booties and gloves. It claims that most of the radioactivity in low-level waste "...fades away to natural background levels in months or years."[15] In contrast, many kinds of toxic wastes remain dangerous forever. Our society presently disposes of some 200 million tons of toxic waste each year, beside which the low-level nuclear waste problem is tiny. The NEI insists that "...engineers have the knowledge and ability to design and build facilities to contain the waste safely for the required period and longer." The NEI also asserts that many disposal options are available for different local environmental conditions.[16]

The failures of previous disposal sites to contain the nuclear waste buried there for even 10 years causes many people to question the wisdom of the current disposal plans. In 1975, the West Valley, New York, waste site was closed after radioactive water burst through the landfill cover. It flowed into streams that feed Lake Erie, a source of drinking water for hundreds of thousands of people. The Maxey Flats site was closed in 1977 when plutonium was found in neighboring property and nearby streams. Due to contamination concerns, the residents of Maxey Flats no longer drink water from local wells.[17]

Activists note that NRC regulations permit disposal sites to leak and deliver a radiation dose as high as 25 millirems per year to people living nearby.[18] That amounts to a dose 25 times higher than what Americans are exposed to from nuclear power plants. For

When used fuel—nuclear waste—is removed from a reactor, it is stored in a large pool of water, where it will cool for at least 10 years.

opponents of nuclear power, this information confirms what they have long feared: the NRC dose standards were set high—not for power plants, but to accommodate the dirtier back end of the fuel cycle. Regulators knew, activists charge, that more radiation would escape into the environment from these disposal sites.

The NEI is confident that modern disposal technology assures greater safety than in the past. It also emphasizes that, in practice, public exposures from low-level waste facilities are substantially lower than the NRC limits.[19] Opponents are not reassured. They note that the high cost of disposal is already leading to a trend called Below Regulatory Concern (BRC). Currently, the NRC requires that all nuclear waste be accounted for and sent to licensed nuclear waste dumps. According to the BRC theory, a certain level of radioactivity is safe enough to be ignored. If adopted, this practice would allow nearly 30 percent of radioactive waste to be disposed of as ordinary garbage—to be put down drains, into landfills, or even recycled.[20] Maine state legislator Maria Holt, a strong opponent of the plan, noted that the combined impact of 20,000 licensees treating radioactive waste as if it were ordinary garbage could be devastating. The Environmental Protection Agency estimates as many as one person in every ten thousand could die as the radiation moves into our everyday environment.[21]

In sharp contrast to what current law allows and what NEI believes is possible, SECC recommends disposal be allowed for only those nuclear wastes with a hazardous life of less than 60 years.[22] The rest, Resnikoff proposed, should be stored at reactor sites until better technology is developed. Some other people—including antinuclear activists, NIMBYs, and industry officials—say that dozens of new dumps aren't needed. We should pick one or two sites, preferably in a desert region. The Nevada atomic bomb test range frequently comes up in discussions of this option. It is already contaminated with radiation and is sometimes called a "national sacrifice zone."

NEI, however, believes new disposal sites are needed, despite the fact that Barnwell will remain open indefinitely. Because of its monopoly, Barnwell is able to charge very high rates. New sites are necessary to promote competition, which could lead to cheaper disposal costs. Otherwise, the possibility remains that the high expense of disposal will bring an end to many of the beneficial activities that generate nuclear waste.[23]

HIGH-LEVEL NUCLEAR WASTE

The issue of low-level radioactive waste disposal continues to create controversy throughout the United States, yet it represents only 1 percent of the total hazard associated with nuclear waste.[24] The other 99 percent is high-level nuclear waste, which consists of spent fuel rods from nuclear reactors and wastes left over from the military building nuclear weapons. About 30,000 metric tons of spent fuel rods are being stored at 110 nuclear plants, and approximately 340,000 cubic meters of military wastes are being stored at three defense facilities: the Savannah River Plant in South Carolina, the Idaho National Engineering Laboratory, and the Hanford Reservation in the state of Washington.[25]

Before a fuel rod goes into a reactor, it can be handled with little danger because uranium 235 is only slightly radioactive. Once the rod has been through a fission cycle of 36–54 months, however, it contains fission products that are so intensely radioactive that anyone coming near it would instantly receive a deadly dose. To prevent this, spent fuel is always kept underwater in special storage pools at nuclear plants.

Nuclear power proponents emphasize the small volume and manageability of spent fuel. All the high-level waste from the past 30 years would only cover a football field to a depth of about nine feet.[26] The NEI says spent fuel is safely stored at nuclear plants in pools or in dry casks made of concrete and steel, and "...an NRC study indicated that spent fuel could be stored at plants for 100 years or more without adverse health or safety consequences...."[27] The American Nuclear Society stresses that "the nuclear waste generated in providing the electricity used by an individual in his entire lifetime is the same size as 100 aspirin tablets."[28]

The extremely compact nature of nuclear fuel has always been one of its attractions. (One ounce of uranium provides as much energy as a ton of coal.) The SECC contends, however, that compactness has little to do with the dangers high-level waste presents. The council noted a 1983 report by the National Academy of Sciences that determined "it would take 3 million years for spent fuel to decay to the point of posing the same level of risk as the uranium ore from which it came."[29] Marvin Resnikoff said that "...to dilute it [those 100 aspirin tablets] to the level considered by the nuclear power industry to be safe for disposal . . . requires 1 billion gallons of water."[30]

ARE WE READY TO THROW IT AWAY?

Keeping high-level nuclear waste from dissolving in water or being spewed into the environment by a volcano or earthquake is the aim of a government effort at Yucca Mountain in Nevada. The Department of Energy is conducting studies to see if the mountain can be a repository for the spent fuel. The NEI claims that most scientists agree that Yucca Mountain is the best site for high-level waste disposal. Dozens of expert committees, such as the American Physical Society, the International

A view of Yucca Mountain in Nevada

Atomic Energy Agency, the U.S. Environmental Protection Agency, and the National Research Council, have confirmed again and again that geologic isolation is the safest long-term option for dealing with high-level waste.[31]

According to the NEI, Yucca Mountain has no volcanoes, no recent history of earthquakes, and the area gets very little rain. Scientists need to be sure that the mountain won't erode away, and the water table won't shift and carry waste to the surface or into drinking water supplies. The spent fuel will be sealed inside a metal container, placed in a lined borehole, surrounded with filler, then sealed with a plug. At Yucca Mountain, the borehole will be in a tunnel 1,000 feet below the surface and 800 feet above the water table. It will remain there for ten thousand years if scientists have correctly evaluated the dryness and stability of Yucca Mountain.

In 1987 Congress chose Yucca Mountain as a possible disposal site. Since then, Yucca Mountain has received detailed scientific study. Yet many people believe this choice represents a failure of the Nuclear Waste Policy Act of 1982. As a measure of political fairness, the act intended the Department of Energy to choose several potential repository sites in the northeast and western parts of the United States. However, widespread public opposition in the heavily populated northeast led to a suspension of that search. In 1987 Congress suspended all other site studies except for Yucca Mountain.

The NEI said this decision occurred because "Congress determined that it was not cost-effective to characterize [study]..." several sites at once.[32] The people of Nevada, and many other critics of the disposal process, have a more colorful way of describing that congressional decision. They call it the "Stick-It-To-Nevada" bill.

Nevada governor Bob Miller believes the nuclear industry is sacrificing democracy "...in its desperate struggle for survival and revival."[33] Nevada has no nuclear plants, and Miller says, "The large majority of Nevadans strongly object to having the risks of the nation's commercial nuclear waste imposed upon them."[34] He insists that Yucca Mountain is subject to earthquakes, volcanoes, and ground water intrusions. In a rush to further the nuclear industry's interests, he believes, democracy and science are suffering. The NEI disagrees completely. It believes deep geological disposal is safe, and that Yucca Mountain has been delayed by NIMBYs and politicians like Miller.

Critics do not think politics alone can explain the difficulties of finding a site and building a safe facility to store high-level waste. They doubt that we have the technology to put spent fuel in the ground and guarantee that it stays put for 10,000 years—a period longer than all recorded human history. The SECC notes that high-level waste tanks at the Hanford, Washington, weapons complex were designed to last 50–100 years. They corroded in only 10 years, allowing nearly 422,000 gallons of radioactive liquid to leak into the soil.[35]

The first plan to dispose of high-level waste involved a salt mine in Lyons, Kansas. After eight years of study, the government was about to begin disposal activities when the Kansas Geologic Survey discovered

The Carey salt mines in Lyons, Kansas, had been proposed as a disposal site for high-level nuclear waste until geologists found it unsuitable. One of many caverns is shown above.

the site "looked like a piece of Swiss cheese."[36] Despite years of intense study, scientists had completely missed the fact that old bore holes existed all around the salt mine.

Yucca Mountain itself is 12 years behind schedule. The nuclear industry wants the government to set up an "interim" waste storage site at Yucca Mountain so that the waste in power plant pools can be moved. Without interim storage, each nuclear plant will have to spend millions of dollars to build new on-site storage facilities. In addition, on-site storage in dry casks, such as the one at the Prairie Island plant in Minnesota, prompted public concern and created a focal point for antinuclear activism.[37] Samuel Skinner is president of a company that operates 12 nuclear plants, and he is a member of NEI's Government Relations Advisory Committee. In testimony before Congress, Skinner said the federal government had made a promise to the industry at the start of the atomic age and had an "unconditional obligation" to take the industry's nuclear waste because "the industry is not prepared to wait any longer. No more empty talk. No more broken promises. No more delays and no more business as usual."[38]

Despite Skinner's urging to "get the job done," the fact remains that 50 years into the nuclear era, no one in the world has committed nuclear waste to permanent disposal. NEI spokesperson Scott Peters believes we have a responsibility to dispose of the waste. He asks, "If not us, who? If not now, when?"

Antinuclear activists believe we have created a million-year hazard for a few decades of electricity. According to these activists, current plans to place a storage site at Yucca Mountain will merely allow the nuclear industry to keep creating this hazard while ignoring our lack of a real solution. Many activists support a call for the president and Congress to establish a blue-ribbon panel to rethink our entire approach to the problem of nuclear waste. According to SECC, the amount of nuclear waste will triple before this generation of 110 plants retires. The council wants to stop the ongoing production of waste by shutting down nuclear plants. Although this plan wouldn't solve the problem of existing waste, it would at least contain the problem for future generations.

Is it our responsibility to do the best we can with current scientific knowledge and technology, or should we wait until better technology comes along? The wrong decision could endanger future generations for thousands of years.

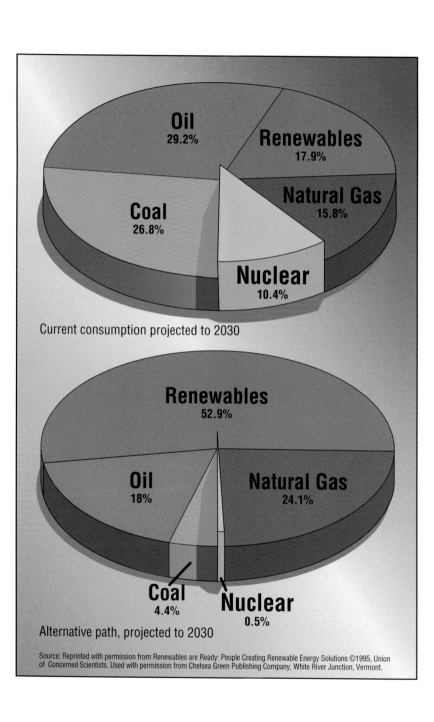

Oil
29.2%

Renewables
17.9%

Coal
26.8%

Natural Gas
15.8%

Nuclear
10.4%

Current consumption projected to 2030

Renewables
52.9%

Oil
18%

Natural Gas
24.1%

Coal
4.4%

Nuclear
0.5%

Alternative path, projected to 2030

TWO PATHS

According to an article in the Orlando *Sentinel,* Donald Hertzmark "...aligns himself with those who feel that...Americans must rid their minds of the near-disastrous and disastrous nuclear mishaps" at Three Mile Island and Chernobyl and get on with the task of building new nuclear reactors.[1] Hertzmark, a member of NEI's speakers' bureau and an international economics consultant, believes the reasons for building new nuclear plants have become more urgent as the world's growing population competes for limited fossil fuels. Nuclear power must be a part of the equation if the United States is to attain energy independence and continue to grow economically.

Amory Lovins is cofounder of the Rocky Mountain Institute in Old Snowmass, Colorado, where he is director of research in resource policy. In contrast to NEI's claim that we need more power plants, Lovins has argued since 1976 that the low-cost, low-risk way to meet future energy needs is not by generating megawatts, but by saving "negawatts." Negawatts is a term he coined to describe energy saved from wasteful practices and appliances.[2] For example, a new, energy-efficient, 25-watt fluorescent lightbulb produces the same amount of light as an old-fashioned 100-watt incandescent bulb. Replacing the old bulb with the new one saves 75 watts of electricity. These negawatts can either power three new 25-watt bulbs, or reduce the need for generating that 75 watts altogether. On a large scale, the potential for saving power is enormous, which led Lovins to conclude that no new power plants of any type are needed. A combination of efficiency and renewable energy could meet our energy needs in a strategy he called the "soft path."[3]

The "soft path" emphasizes small-scale solutions, such as using energy more wisely and tapping sun, wind, and water as sources of energy. The "hard path" represents the old strategy of meeting demands for energy by supplying new power—typically with huge coal and nuclear plants.

In an article titled "The Irrelevance of Nuclear Power," Lovins wrote, "Since nuclear power is unnecessary and uneconomic, we needn't debate whether it is safe, clean, or peaceful; we simply shouldn't buy it."[4]

In contrast, NEI looks at trends showing electric demand keeping step with economic growth and rising in all sectors of the economy. It comes to a different conclusion: Even if efficiency can do all that enthusiasts like Lovins claim—which NEI doesn't believe—we will still need to build new power plants. In fact, NEI says we must maximize the "use of all practical renewable energy supplies as well as conventional power sources—coal, nuclear, oil, and gas."[5]

But SECC says industry officials always paint

gloomy scenarios of a future without nuclear plants. In 1976, industry officials raised the specter of electricity shortages and blackouts unless all planned nuclear plants were built, but these things never occurred despite the cancellation of over 100 reactors.[6] Lovins noted that worldwide "nuclear capacity in 2000 will be only 2–3 percent of official 1974 projections for developing and at most 4–8 percent for developed countries."[7]

Skeptics of the "soft path" often ask whether the American public is willing to adopt the policies needed for efficiency, or is ready to accept changes in their energy-abundant lifestyles. President Jimmy Carter, wearing a sweater and advising the nation to turn down the thermostat, was a key image for promoting energy conservation during the 1970s. In the 1990s,

During his term in office (1977-1981), President Jimmy Carter urged people to conserve energy.

after a decade of cheap oil, many Americans seem to
have rejected energy conservation austerities. They are
buying more gas-guzzling cars and raising speed limits.
The United States is 50 percent dependent on foreign
oil—a higher level than when the Arab oil embargo occurred in the 1970s and early 1980s.

Energy efficiency is not the same as conservation,
Lovins argues. *Efficiency* means doing the same thing
with less energy. The food is just as cold in an efficient
refrigerator, yet the best models use 10 times less energy than inefficient ones. Lovins's own home in Colorado has sacrificed none of the modern comforts, yet
his electric bill is $5 per month (one-tenth that of the
average home), and the heat comes almost entirely

*Solar-powered homes, like the one above, are energy-
efficient.*

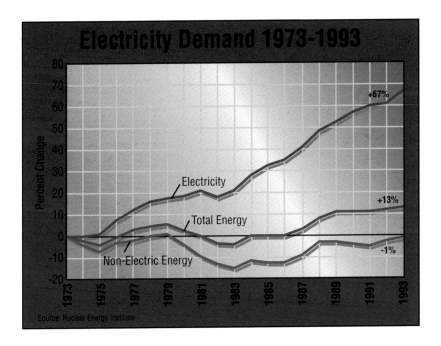

Electricity Demand 1973-1993

+67%

+13%

Electricity

Total Energy

Non-Electric Energy

-1%

Percent Change

Source: Nuclear Energy Institute

from solar energy. SECC says that rejecting the efficiency path also puts the United States at a disadvantage in the global economy. Japanese and German industries, for example, use half as much energy as the United States does to make an industrial product.[8] This savings results in cheaper prices for their goods. By simply using better motors, lights, and machines, German and Japanese industries do not have to spend money on energy that "goes down the drain."

The NEI, however, thinks the future will require ever increasing supplies of electrical energy. NEI believes Lovins and SECC are wrong about the ability of energy efficiency and renewables to meet future energy demands. They note that by the year 2000, 36 percent of America's large power plants will be too old to operate.

While Lovins might claim that we will not miss them, NEI is certain that we will have to replace them.

NEI has developed a strategy for building new nuclear power plants to meet that demand. The institute's strategic plan involves securing one-step licensing of standardized advanced reactors and resolution of the nuclear waste problem. According to NEI, almost every nuclear plant in the United States is a one-of-a-kind design. In the past, this situation contributed to high costs, delays, and overregulation. The new reactors will be standardized, says NEI: "Since today's nuclear plants were designed and built, there have been tremendous strides in many technological areas. The 'evolutionary' designs optimize the light water reactor, producing a plant that is simpler, easier to operate and maintain, and costs less to build. Safety studies indicate that these designs will be able to meet safety goals that are more than 100 times greater than those of current plants."[9]

France, where 75 percent of the electricity is produced by nuclear power, uses standard designs. The cost of their plants is among the lowest in the world, and construction takes only 5 to 6 years. In the United States, construction has taken 10 to 12 years. Opponents of NEI's plan point out that the French government controls nuclear power, not private utilities as in the U.S. free-market system. Public participation in France is minimal. As the U.S. nuclear industry tries to move toward such models, SECC argues that democracy suffers. A case in point is the one-step licensing rule, which denies the public a hearing before a nuclear plant goes into operation.

A nuclear energy plant in Cadarache, France, near the Mediterranean Sea

Before 1992, the public could question the construction of a nuclear plant. Even after the plant was built (10 or 15 years later), the public could still question whether it should go into operation. According to SECC, this second hearing led to improved safety in at least 10 plants.[10] But Richard Myers, an NEI vice-president, sees it differently. The old way led to costly delays, but now "...all major issues...will be settled before a company starts to build a nuclear power plant....Everybody wins."[11]

The big prize, the industry hopes, will be the first new order for a nuclear plant in more than 20 years. Albert B. Reynolds, professor of nuclear engineering at the University of Virginia said, "By the year 2000, we

Professor Albert B. Reynolds, of the University of Virginia, predicts an increased reliance on nuclear energy.

should see the reemergence of nuclear energy as the most promising power source for generating electricity. Cheaper than gas and safer than coal, with a near-infinite fuel supply, nuclear power's time has come. Again."[12]

"We've heard that promise before," SECC responded in a recent report on nuclear power costs. SECC concluded that nuclear electricity is the most expensive conventional electricity, and it is an economic burden to ratepayers. According to SECC, electricity customers who get power from nuclear plants pay $200 more per year than customers who buy from conventional sources.[13] Billions of dollars of new investment will also be needed to revive nuclear power. SECC contrasted Reynolds's confidence with the words of Dan Scotto, an electric utility analyst, who said, "I think the technology [for nuclear power] has for all intents and purposes become obsolete. Other fuel types are considerably less expensive."[14]

But NEI points to favorable indicators for a nuclear revival that include successful enactment of one-step licensing—which President George Bush signed into law on October 24, 1991—and $200 million in government funds for design work on two advanced reactors.[15] SECC notes, however, that the last three nuclear plants under construction were canceled, and a survey of utility executives revealed that three-quarters of them believed their companies would never order another nuclear plant.[16]

In a world of cheap oil, both sides do agree that government support is needed to plan ahead for a time of fossil fuel scarcity. SECC believes the government should put taxpayers' money where it thinks the real payoff is. In 1994, according to SECC, energy efficiency saved enough waste energy to meet 30 percent of new energy needs.[17] Nuclear power supplied only 6 percent. Nevertheless, in that same year the nuclear industry received 60 percent of all government research and development funds, while efficiency technologies received only 6 percent. SECC says a 1992 study by Kamanoff Energy Associates showed that the federal government had invested $97 billion in the nuclear industry since 1950.[18] The result has been expensive electricity and an unsolved waste problem. SECC cited a 1987 Department of Energy report that determined renewable energy could supply all our energy needs. SECC also argues that after 50 years of subsidizing nuclear power, the government ought to try supporting renewable energy.[19]

NEI claims that we must keep all our energy options open, and it points out that when the government did spend money intended to encourage renewable energy, the results were not as good as predicted. By the end of 1988, renewable energy projects had met only 35 percent of projections, due primarily to an inability to compete with cheap natural gas.[20]

A FAUSTIAN BARGAIN

Is nuclear power a "relic of a failed energy policy," as SECC executive director Scott Denman believes, or is it ready for a "reemergence . . . as the most promising power source for electricity," as Professor Reynolds and the NEI believe?

Have the failures and setbacks simply been the growing pains of a new technology—a period of shaking out the bugs, educating the public to the benefits, and calming fears? Or, after 50 years, has the public decided that the perils are not worth the prize? Was the promise of power "too cheap to meter" too ambitious? As long as the United States remains a democracy, we must all decide together.

In 1972, Dr. Alvin Weinberg, then Director of the Oak Ridge National Laboratory, first said, "We nuclear people have made a Faustian bargain with society."[21] Faust was a legendary alchemist who sold his soul to the devil for unlimited power. Weinberg, a lifelong believer in nuclear energy, explained what he meant by "Faustian bargain" in an article about Chernobyl. He wrote, "Humans, in opting for nuclear energy, must pay the price of extraordinary technical vigilance for the energy they derive from nuclear fission if they are to avoid serious trouble."[22]

In the 1940s, our nation signed that bargain. The nuclear industry wants our society to renew it with a commitment to a new generation of reactors. Their goal is a new reactor order by the year 2000.

How will you feel if it's in your community?

Resources to Contact

American Nuclear Society (ANS)
North Kensington Avenue
LaGrange Park, IL 60525
(312) 352-6611

National Energy Information Center
    (NEIC)
Energy Information Administration
U.S. Department of Energy, EI-231
Forrestal Building, Room IF 048
Washington, DC 20585
(202) 586-8800

Nuclear Energy Institute (NEI)
Suite 400
1776 I Street, NW
Washington, DC 20006-3708
(202) 739-8000

Nuclear Information and Resource
    Service (NIRS)
1424 16th Street, Suite 601
Washington, DC 20036
(202) 328-0002

Rocky Mountain Institute (RMI)
1739 Snowmass Creek Road
Snowmass, Colorado, 81654-9199
(970) 927-3851

Safe Energy Communication Council
    (SECC)
Suite 805
1717 Massachusetts Avenue, NW
Washington, DC 20036
(202) 483-8491

Your local Electric Power Utility
    Company

Your State Energy Office

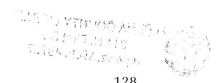

ENDNOTES

CHAPTER 1. PROMISE OR PERIL?

[1]*Nuclear Power Plant Safety* (Washington, D.C.: U.S. Council for Energy Awareness, November 1991), newsletter.

[2]Ibid.

[3]Janet S. Wager, "How Much Are Fossil Fuels Really Costing You?" *Nucleus,* Summer 1993, 6.

[4]Ibid., 4.

[5]*1993 Annual Energy Review,* U.S. Energy Information Administration.

[6]Wager, 6.

[7]Mike Edwards, "Living with the Monster: Chernobyl," *National Geographic,* August 1994, 104.

[8]Vermont Yankee press release, January 31, 1989.

[9]Direct communication, December 11, 1995.

[10]SECC press release, September 19, 1995.

[11]Direct communication, December 13, 1995.

[12]Direct communication, December 13, 1995.

CHAPTER 3. ENERGY AND CIVILIZATION

[1]Glen Seaborg and Corliss Seaborg, *Man and Atom* (New York: E.P. Dutton, 1971), 20–21.

[2]David O. Woodbury, *Atoms for Peace* (New York: Dodd Mead & Co., 1955), 102.

[3]*Sustainable Energy Strategy* (Washington, D.C.: U.S. Department of Energy, July 1995), 9.

[4]Woodbury, 102.

[5]*The Atomic Age Opens,* ed. Donald Porter Geddes and Gerald Wendt (New York: Pocket Books, USA, 1945), 186.

[6]Thomas R. Wellnitz, "An Energy Activity Sampler," *Social Education,* (Washington, D.C.: National Council for the Social Studies, January 1992), 26.

[7]"Electricity Trends," *Info*Bank: A series of fact sheets and data on energy, electricity and nuclear power (Washington, D.C.: Nuclear Energy Institute, July 1994), 1.

[8]Donella Meadows, Dennis Meadows, and Randers Jorgen, *Beyond the Limits* (Post Mills, Vermont: Chelsea Green, 1992), 7.

[9]Ibid., 68.

[10]Ibid., 78.

[11]WEC Commission, *Energy for Tomorrow's World* (New York: St. Martins Press, 1993), 43.

[12]Ibid., 41.

[13]*Funk & Wagnalls World Almanac.*

[14]WEC Commission, 40.

[15]Meadows, 66.

[16]*Nuclear Power: Both Sides,* ed. Michio Kaku and Jennifer Trainer (New York: W. W. Norton, 1982), 2

[17]Corinne Browne and Robert Munroe, *Time Bomb: Understanding the Threat of Nuclear Power* (New York: William Morrow & Co., 1981), 37.

[18]Ibid.

[19]Ibid.

[20]*The Atomic Age Opens,* 178.

[21]Leslie Lamkin, "A Triumph of the Human Spirit," *Nuclear Industry,* Fourth Quarter, 1992, 9.

[22]Ibid.

[23]David E. Lilienthal, *Atomic Energy: A New Start* (New York: Harper & Row, 1980), 1.

[24]Ibid., 2.

[25]Ibid.

[26]Daniel Ford, *The Cult of the Atom: The Secret Papers of the Atomic Energy Commission* (New York: Simon and Schuster, 1982), 41.

[27]Browne and Munroe, 86.

[28]Lamkin, 13.

[29]Ford, 62.

[30]Ibid., 63.

[31]"2042 . . . Here We Come," *Nuclear Industry,* Fourth Quarter 1992, 35.

[32]Lilienthal, 29.

CHAPTER 4. POLLUTION

[1]Larry Tye, "Edison Assails State's Report on Cancer Rate near Pilgrim," *Boston Globe,* December 29, 1988.

[2]Ibid.

[3]"Leukemia Risk Found near Pilgrim in 1972–1979," *Boston Globe,* October 10, 1990.

[4]"Radiation," *Info*Bank, 9.

[5]Ibid., 1.

[6]Kay Drey, "Nuclear Power's Dirty Secret," SECC *Viewpoint,* 1990.

[7]Martin Resnikoff, "Low-Level Radioactive Waste," *Mythbuster #8* (Washington, D.C.: Safe Energy Communication Council, 1992), 11.

[8]Alex Antypas, "The Greenhouse Effect," *Mythbuster #4* (Washington, D.C.: Safe Energy Communication Council, 1989), 5.

[9]Kaku and Trainer, 162.

[10]Bernard L. Cohen, *Before It's Too Late, A Scientist's Case for Nuclear Energy* (New York: Plenum Press, 1983), 98.

[11]Committee for Nuclear Responsibility, "Back to Living in the Caves?" (undated leaflet).

[12]Joseph Kriesberg, "Foreign Oil Dependence," *Mythbuster #3* (Washington, D.C.: Safe Energy Communication Council, 1988), 7.

[13]"Utility Fuel Displacements," *Info*Bank, July 1994.

[14]Scott Denman, "Nuclear Power No Solution to Foreign Oil Dependence," SECC *Viewpoint,* 1988.

[15]Ibid.

[16]Antypas, 6.

[17]Karl Grossman, *Coverup: What You Are Not Supposed to Know about Nuclear Power* (New York: The Permanent Press, 1980), 80.

[18]Kaku and Trainer, 112.

[19]Cohen, 13.

[20]Resnikoff, 11.

[21]"Reducing Radon Risks," EPA pamphlet, 1992 (Washington, D.C.: U.S. Government Printing Office).

[22]Anita Warren, "Radiation Phobia, It Could Be Hazardous to Your Health," *Nuclear Industry*, Third Quarter 1991.

[23]Ibid.

[24]John W. Gofman and Arthur R. Tamplin, *Poisoned Power* (Emmaus, Pennsylvania: Rodale Press, 1979), 97.

[25]Resnikoff, 11.

[26]Gofman and Tamplin, 97.

[27]"Radiation," *Info*Bank, 7.

CHAPTER 5. NUCLEAR SAFETY

[1]*Nuclear Power Plant Safety*, U.S. Council for Energy Awareness, November 1991.

[2]Martin Gelfand, "Nuclear Reactor Safety," *Mythbuster #7* (Washington, D.C.: Safe Energy Communication Council, 1992), 5.

[3]Ibid.

[4]Ibid., 9

[5]Scott Denman, "Three Mile Island: The Accident Continues," SECC *Viewpoint*, 1989.

[6]Scott Denman, "A Lesson Lost? America's Response to Chernobyl," SECC *Viewpoint*, 1989

[7]U.S. Council for Energy Awareness, *Nuclear Power Plant Safety*, November 1991.

[8]Jeff Donn, "Debate Escalates over Rowe Plant, *Brattleboro Reformer*, July 29, 1991.

[9]Cohen, 132.

[10]Donn, "Debate Escalates."

[11]Theresa M. Maggio, "NRC Hears Impassioned Pleas to Close Rowe," *Brattleboro Reformer*, July 24, 1991.

[12]"Residents Skeptical about Yankee Rowe Emergency Plans," *Brattleboro Reformer*, September 23, 1991.

[13]Michael Daley, "Brittle Defense," *On the Watch*, New England Coalition on Nuclear Pollution, 3 (newsletter).

[14]Maggio, "Impassioned Pleas."

[15]Daley, "Brittle Defense."

[16]Kaku and Trainer, 86.

[17]Ibid., 90.

[18]Ibid., 89, 97.

[19]Ibid., 104.

[20]Ibid., 106.

[21]Ibid., 107.

[22]Ibid., 97-98.

[23]Christopher Flavin, *Worldwatch Paper 75, Reassessing Nuclear Power: The Fallout from Chernobyl* (Washington, D.C.: Worldwatch Institute, 1987), 40.

[24]Gofman, 163.

[25]Kaku and Trainer, 84.

[26]Robert Del Tredici, *The People of Three Mile Island* (San Francisco: Sierra Club Books, 1980), 31.

[27]"Radiation," *Info*Bank, 2.

[28]Ibid., TMI-2, 2.

[29]Tredici, 33.

[30]Lamkin, 19.

[31]Kaku and Trainer, 105.

[32]Ibid., 25.

[33]Daniel Martin, *Three Mile Island: Prologue or Epilogue?* (Cambridge, Massachusetts: Ballinger Publishing Co., 1980), 208.

[34]*Info*Bank, TMI-2, 7.

[35]Ibid.

[36]Flavin, 51.

[37]David Kraft, "Safety Last at Nuclear Power Plants," SECC *Viewpoint*, 1989.

[38]Denman, "Lesson Lost."

[39]Tredici, 48

[40]Anna Gyorgy et al., *No Nukes: Everyone's Guide to Nuclear Power* (Boston: South End Press, 1979), 415.

[41]Telephone interview with Dr. Judith Johnsrud, December 1995.

[42]Alla Yaroshinkskaya, *Chernobyl: The Forbidden Truth,* trans. Michele Kahn and Julia Sallabank (Lincoln, Nebraska: University of Nebraska Press, 1995), 83.

[43]Ibid.

[44]British Broadcasting Corporation, "World Report," *WAMC Midday Magazine,* October 26, 1995,

[45]"Chernobyl Accident," *Info*Bank, 1.

[46]Frederick Seitz, "Must We Have Nuclear Power?" *Readers Digest,* August, 1990.

[47]W. Jack Bowen, "Energy for Tomorrow's World," *World Energy Council Journal.*

[48]Tredici, 31.

[49]Telephone interview with Debbie Katz, December 1995.

[50]Hunt Williams, "Nuclear Images: Thanks for the Memories, Godzilla," *Nuclear Industry,* Fourth Quarter, 1992, 31.

[51]Ibid.

[52]Telephone interview with Diana Sidebotham, December 1995.

[53]Seitz, 4.

[54]Christopher Flavin, Worldwatch Paper 57, *Nuclear Power: The Market Test* (Washington, D.C.:Worldwatch Institute, 1987), 24.

[55]Gyorgy et al, 400.

[56]Gelfand, 2.

[57]Ibid.

[58]"Rowe Plant Shut Down," *Brattleboro Reformer,* October 2, 1991.

[59]Flavin, *Market Test,* 47.

[60]"Voter Initiatives," *Info*Bank.

CHAPTER 6. WASTE DISPOSAL

[1]Emily Carmain, "Barnwell's Good Neighbor," *Nuclear Industry,* Fourth Quarter 1992, 42.

[2]Marvin Resnikoff, *Living Without Landfills* (New York: Radioactive Waste Campaign, 1987), 40.

[3]Carmain, 42.

[4]"Low-Level Waste," *Info*Bank, July 1994, 5.

[5]Resnikoff, *Landfills,* 33.

[6]"Low-Level Waste," *Info*Bank, July 1994, 7.

[7]Carmain, 42.

[8]Carmain, 45.

[9]Carmain, 43.

[10]"Low-Level Waste," 9.

[11]*Nuclear Industry,* Fourth Quarter 1993, advertisement, 49.

[12]Resnikoff, *Landfills,* 5.

[13]*Nuclear Industry,* Fourth Quarter 1993, advertisement, 49.

[14]Ibid.

[15]"Low-Level Waste," 7.

[16]Ibid., 6.

[17]Resnikoff, *Mythbuster #8,* 3.

[18]"Low-Level Waste," 7.

[19]Ibid.

[20]Maria Holt and Joan Bozer, "Radioactive Waste: Coming Soon to a Landfill near You," SECC *Viewpoint,* 1990.

[21]Ibid.

[22]Resnikoff, *Mythbuster #8,* 13.

[23]*Low-Level Waste: Beyond Barnwell* (Washington, D.C.:U.S. Committee for Energy Awareness [now NEI], 1992), pamphlet.

[24]"Low-Level Waste," 7.

[25]James Lawson, "Nuclear Waste Disposal," *Mythbuster #2* (Washington, D.C.: Safe Energy Communication Council, 1988), 2.

[26]"High-Level Waste," *Info*Bank, July 1994, 1.

[27]Ibid., 2.

[28]Lawson, 1.

[29]Ibid., 3.

[30]Resnikoff, "Nuclear Wastes—The Myths and the Realities," *Sierra,* July/August 1980, 1.

[31]Alice Camp, "Of Mountains and Molehills," *Nuclear Industry,* Third Quarter 1991.

[32]"High-Level Waste," 3.

[33]Bob Miller, "Surrendering Democracy to Nuclear Waste," SECC *Viewpoint,* 1992.

[34]Ibid.

[35]Lawson, 4.

[36]Ibid.

[37]Sinner, Samuel, "Nuclear Waste: No More Excuses," *Nuclear Industry,* First Quarter 1995, 6.

[38]Ibid.

CHAPTER 7. TWO PATHS

[1]Dick Marlowe, "Experts Say U.S. Needs to Exert More Energy Toward an Atomic Future," *Orlando Sentinel,* December 16, 1994.

[2]Art Kleiner, "Amory Lovins: Two Times," *Popular Science,* July 1992, 79.

[3]Ibid.

[4]Amory Lovins, "The Irrelevance of Nuclear Power," *Denver Post,* June 1986.

[5]"Renewables," *Info*Bank, 1.

[6]Gelfand, Martin, "Nuclear Power Economics," *Mythbuster #9* (Washington, D.C.: Safe Energy Communications Council, Fall 1995), 3.

[7]Lovins.

[8]Alex Antypas, "The Greenhouse Effect," *Mythbuster #4* (Washington, D.C.: Safe Energy Communications Council, Fall 1995),3.

[9]"Advanced," *Info*Bank, 3.

[10]Gelfand, *Mythbuster #7,* 12.

[11] Richard Myers, *Nuclear Industry,* Fourth Quarter 1992, editorial.

[12]Gelfand, *Mythbuster #9,* 7

[13]Ibid.

[14]Ibid.

[15]"Advanced," *Info*Bank, 6.

[16]Gelfand, *Mythbuster #9,* 8–9.

[17]Ibid., 10.

[18]Ibid.

[19]Kriesberg, 8.

[20]"Renewables," *Info*Bank, 2.

[21]Grossman, 14.

[22]Alvin Weinberg, "A Nuclear Power Advocate Reflects on Chernobyl," *Bulletin of the Atomic Scientists,* August/September 1986, 57.

Glossary

arms race: a state of affairs in which two adversaries continually build new weapons [and larger armies] in an effort to remain more powerful than their enemy

Atomic Energy Commission (AEC): a federal agency formed by Congress in 1946 to control all aspects of atomic technology. The AEC was given the tasks of developing and promoting atomic power while also insuring it was used safely. This conflict of interest forced Congress in 1974 to separate the AEC into the Nuclear Regulatory Commission and the Department of Energy.

civil disobedience: the deliberate, usually non-violent, violation of a law in order to demonstrate that it is a bad law

"clogging the toilet": a term referring to the anti-nuclear activist tactic of creating opposition to all nuclear waste disposal facilities. The goal is to force nuclear power plants to close by overwhelming their on-site storage facilities with waste.

cold war: a period of high tension that began after World War II between the nuclear superpowers of the United States and the Soviet Union

compact: a practice authorized by Congress in which a number of states may join together to create a single low-level nuclear waste disposal facility for all the members of the compact

emergency core cooling system (ECCS): the pumps, pipes, and control equipment designed to supply water to the reactor in the event of a loss of the normal cooling water supply. Its operation is essential to prevent a meltdown

emergency shutdown or SCRAM: the immediate insertion of the control rods designed to occur within seconds after the protective systems of a reactor detect a dangerous condition

energy conservation: the practice of eliminating the use of energy you previously thought was essential. For example, deciding to make only one automobile trip to the grocery store each week instead of five is a choice that conserves energy.

energy efficiency: a practice or a technology that reduces the amount of energy needed to achieve the same result. For example, insulating an old house will reduce the amount of oil burned in the furnace, yet the house will be just as, if not more, comfortable than it was before.

Environmental Protection Agency (EPA): the federal agency charged with developing regulations that protect the environment and correcting the harm caused by past practices

fossil fuel: any naturally occurring organic fuel, such as petroleum, coal, and natural gas

global warming: the scientific theory that says the earth's overall temperature is increasing due to a buildup of various greenhouse gases being released by human activity— principally, carbon dioxide

grassroots: a term used to describe an upwelling of action that starts with the ordinary citizen, usually in opposition to some undesirable activity authorized by government or other organizations

hearings: formal meetings that are similar to a trial in a court. A hearing board acts as the judge, while interveners and their witnesses argue for or against the proposed action.

license: an extremely complex document granted to the owner of a nuclear power plant by the NRC. It gives the owner the right to operate a reactor and describes the conditions that the owner must meet to ensure that the reactor operates safely.

meltdown: a nuclear accident in which the core of nuclear fuel melts due to overheating. A meltdown carries the potential of releasing enormous amounts of radioactivity into the environment.

Mutual Assured Destruction (MAD): a nuclear weapons policy based on the idea that no matter how badly one nation damages another in a nuclear war, that nation will have enough weapons left over to totally destroy its attacker

NIMBY or "not in my backyard": a term used to describe the local opposition [often intense] to new facilities that are perceived as undesirable by the community. Typically, NIMBY is used as a slur denoting an emotional, unreasoned opposition.

NRC inspectors: employees of the NRC who work full time at nuclear power plants. They observe, evaluate and report on the adequacy of safety at the nuclear plant where they work.

nuclear fuel cycle: a term referring to all the elements necessary to support the operation of a nuclear reactor. The nuclear fuel cycle includes the mining, processing, shipping, using, reprocessing, and disposing of radioactive material.

Nuclear Regulatory Commission (NRC): a federal agency formed by Congress in 1974 to license, inspect, and enforce the rules for all users of radioactive materials in the United States. The NRC is headed by a five-member commission appointed by the president. It has a staff of over 3,000 employees who conduct regulatory research and develop safety standards, as well as monitor the daily operation of U.S. reactors.

ratepayer: the people who pay the electric bills.

regulations: a body of rules adopted by the Nuclear Regulatory Commission

renewable energy [alternative energy sources, sustainable energy]: energy obtained from sources that are essentially inexhaustible (unlike fossil fuels of which there is a finite supply). Solar, wind, and water energy are examples.

turbine-generator: the combination of a steam turbine and an electrical generator to convert the heat energy of a reactor into electrical power

"watchdog": a term often applied to a citizen, group, or agency that pays close attention to the behavior of industry, business, or government in order to point out abuses, failures, and problems

watt: a measure of electrical energy. Home energy use is measured in kilowatts (thousands of watts), while the energy output of power plants is typically measured in megawatts (millions of watts).

Bibliography

Antypas, Alex. *Mythbuster #4: The Greenhouse Effect.* Washington, D.C.: Safe Energy Communication Council, 1989.

Browne, Corinne, and Robert Munroe. *Time Bomb: Understanding the Threat of Nuclear Power.* New York: William Morrow & Co., 1981.

Camp, Alice. "Of Mountains and Molehills." *Nuclear Industry,* Third Quarter 1991.

Carmain, Emily. "Barnwell's Good Neighbor." *Nuclear Industry,* Fourth Quarter 1992.

Cohen, Bernard L. *Before It's Too Late: A Scientist's Case for Nuclear Energy.* New York: Plenum Press, 1983.

Daley, Michael. "Brittle Defense." *On the Watch.* New England Coalition on Nuclear Pollution, 3 (newsletter).

Denman, Scott. "Nuclear Power No Solution to Foreign Oil Dependence." SECC *Viewpoint,* 1988.

———. "Three Mile Island: The Accident Continues." SECC *Viewpoint,* 1989.

Donn, Jeff. "Debate Escalates over Rowe Plant." *Brattleboro Reformer,* July 29, 1991.

Drey, Kay. "Nuclear Power's Dirty Secret." SECC *Viewpoint,* 1990.

Edwards, Mike. "Living with the Monster: Chernobyl." *National Geographic,* August 1994.

Flavin, Christopher. *Worldwatch Paper 75, Reassessing Nuclear Power: The Fallout from Chernobyl.* Washington, D.C.:Worldwatch Institute, 1987.

Ford, Daniel. *The Cult of the Atom: The Secret Papers of the Atomic Energy Commission.* New York: Simon and Schuster, 1982.

Geddes, Donald Porter, and Gerald Wendt, eds. *The Atomic Age Opens.* New York: Pocket Books, 1945.

Gelfand, Martin. *Mythbuster #7: Nuclear Reactor Safety.* Washington, D.C.: Safe Energy Communication Council, 1992.

———. *Mythbuster #9: Nuclear Power Economics.* Washington, D.C.: Safe Energy Communication Council, 1995.

Gofman, John W., and Arthur R. Tamplin. *Poisoned Power.* rev. ed. Emmaus, Pennsylvania: Rodale Press, 1971, 1979.

Grossman, Karl. *Coverup: What You Are Not Supposed to Know about Nuclear Power.* New York: The Permanent Press, 1980.

Gyorgy, Anna, and friends. *No Nukes: Everyone's Guide to Nuclear Power.* Boston: South End Press, 1979.

Holt, Maria and Joan Bozer. "Radioactive Waste: Coming Soon to a Landfill near You." SECC *Viewpoint,* 1990.

Kaku, Michio, and Jennifer Trainer, eds. *Nuclear Power: Both Sides.* New York: W. W. Norton, 1982.

Kleiner, Art. "Amory Lovins: Two Times." *Popular Science,* July 1992.

Kraft, David. "Safety Last at Nuclear Power Plants." SECC *Viewpoint,* 1989.

Kriesberg, Joseph. *Mythbuster #3: Foreign Oil Dependence.* Washington, D.C.: Safe Energy Communication Council, 1988.

Lamkin, Leslie. "A Triumph of the Human Spirit." *Nuclear Industry,* Fourth Quarter 1992, 5–23.

Lawson, James. *Mythbuster #2: Nuclear Waste Disposal.* Washington, D.C.: Safe Energy Communication Council, 1988.

"Leukemia Risk Found near Pilgrim in 1972–1979," *Boston Globe,* October 10, 1990.

Lilienthal, David E. *Atomic Energy: A New Start.* New York: Harper & Row, 1980.

Lovins, Amory. "The Irrelevance of Nuclear Power." *Denver Post,* June 1986.

Maggio, Theresa M. "NRC Hears Impassioned Pleas to Close Rowe." *Brattleboro Reformer,* July 24, 1991.

Marlowe, Dick. "Experts Say U.S. Needs to Exert More Energy toward an Atomic Future." *Orlando Sentinel,* December 16, 1994.

Martin, Daniel. *Three Mile Island: Prologue or Epilogue?* Cambridge, Massachusetts: Ballinger Publishing, 1980.

Meadows, Donella, Dennis Meadows, and Randers Jorgen. *Beyond the Limits.* Post Mills, Vermont: Chelsea Green, 1992.

Miller, Bob. "Surrendering Democracy to Nuclear Waste." SECC *Viewpoint,* 1992.

Myers, Richard. Editorial. *Nuclear Industry,* Fourth Quarter 1992.

Nuclear Energy Institute. *Info*Bank: A series of fact sheets and data on energy, electricity, and nuclear power. Washington, D.C.: Nuclear Energy Institute.

Nuclear Power Plant Safety. Washington, D.C.: U.S. Council for Energy Awareness, November 1991.

"Residents Skeptical about Yankee Rowe Emergency Plans." *Brattleboro Reformer,* September 23, 1991.

Resnikoff, Marvin. *Living without Landfills.* New York: Radioactive Waste Campaign, 1987.

———. *Mythbuster #8: Low-Level Radioactive Waste.* Washington, D.C.: Safe Energy Communication Council, 1992.

Rickard, Graham. *The Chernobyl Catastrophe.* New York: Bookwright Press,1989.

"Rowe Plant Shut Down." *Brattleboro Reformer,* October 2, 1991.

Seaborg, Glen, and Corliss Seaborg. *Man and Atom.* New York: E. P. Dutton, 1971.

Seitz, Frederick. "Must We Have Nuclear Power?" *Readers Digest,* August 1990.

Sinner, Samuel. "Nuclear Waste: No More Excuses." *Nuclear Industry,* First Quarter.

Tredici, Robert Del. *The People of Three Mile Island.* San Francisco: Sierra Club Books, 1980.

Tye, Larry. "Edison Assails State's Report on Cancer Rate near Pilgrim." *Boston Globe,* December 29, 1988.

United States Department of Energy. *Sustainable Energy Strategy.* n. p.: July 1995.

Wager, Janet S. "How Much Are Fossil Fuels Really Costing You?" *Nucleus,* Summer 1993.

Warren, Anita. "Radiation Phobia, It Could Be Hazardous to Your Health." *Nuclear Industry,* Third Quarter, 1991.

WEC Commission. *Energy for Tomorrow's World.* New York: St. Martins Press, 1993.

Weinberg, Alvin. "A Nuclear Power Advocate Reflects on Chernobyl." *Bulletin of the Atomic Scientists,* August/September 1986.

Wellnitz, Thomas R. "An Energy Activity Sampler." *Social Education.* (Washington D.C.: National Council for the Social Studies, January 1992), 26.

What Do You Know about Low-Level Radioactive Waste? Washington D.C.: U.S. Council for Energy Awareness, n.d.

Williams, Hunt. "Nuclear Images: Thanks for the Memories, Godzilla." *Nuclear Industry,* Fourth Quarter, 1992, 25–33.

Woodbury, David O. *Atoms For Peace.* New York: Dodd Mead, 1955.

Yaroshinkskaya, Alla. *Chernobyl: The Forbidden Truth.* Translated by Michele Kahn and Julia Sallabank. Lincoln, Nebraska: University of Nebraska Press, 1995.

Index

acid rain, 9, 11, 52, 53
air pollution, 10–11, 45, 51, 53–56
American Nuclear Society, 110
American Physical Society, 111
Asselstine, James K., 94–95
atomic bombs, 7–8, 32–36, 37–39, 59, 62
atomic energy, 31–43
Atomic Energy Act, 36, 37, 71, 97
Atomic Energy Commission (AEC), 28, 36–37, 40, 42, 76, 93, 95
atoms, 21–22, 36, 56–57

Beyea, Jan, 78–79, 84
Bird, Ralph, 46–47
Bush, George, 125

cancer, 14–15, 45–48, 49, 59, 62, 64, 99
carbon dioxide (CO2), 10–11, 43, 52, 53–54, 56, 90
Carter, Jimmy, 119
Chernobyl nuclear plant, 12–15; results of meltdown at, 16, 17, 18, 41, 70, 85, 88–89
Citizens' Awareness Network, 91
Citizens Urging Responsible Energy, 45
coal, 7, 9, 10–12, 21, 28, 30, 40, 41, 51–52, 53–54
Cobb, Sidney, 47
Cohen, Bernard, 51–52, 63, 92
Committee for Nuclear Responsibility, 52
Committee on the Biological Effects of Ionizing Radiation, 61
Congress, U. S., 36, 40, 97, 102, 112–115
containment buildings, 14, 18, 21, 25, 68, 69, 82, 83, 85, 88
control rods, 22, 25, 81
control room, 81, 84, 85, 96

cooling systems, 13, 22–23, 25, 57, 69. See also water
core, 8, 14, 22, 57, 73, 81, 82, 85

Denman, Scott, 17–18, 54–55, 85, 126
Department of Energy (DOE), 55, 111, 112, 125

Eisenhower, Dwight D., 38
electricity, 10, 16, 21, 28–30, 117–127
Electric Power Research Institute, 43
electrons, 22, 57
emergency core cooling systems (ECCS), 13, 25, 40, 42, 68, 73, 81
energy, 27–31; atomic, 31–43; efficiency, 119–121; future of, 117–127
Environmental Protection Agency (EPA), 49, 61–62, 108, 112

fallout, 14, 38, 39, 89
Federal Emergency Management Agency (FEMA), 96
Fermi, Enrico, 32, 33
fission, 7, 10, 14, 21, 22, 24–25, 31–32, 37, 57–58, 81, 110
Flavin, Christopher, 94, 97
fossil fuels. See coal; natural gas; oil
fuel assemblies, 22–23
fuel rods, 22, 24, 25, 57, 110

General Electric, 42, 69
generators, 23–24, 74; electric, 8, 80–81; steam, 23
global warming, 11, 19, 42–43, 52, 53, 90
Gofman, John, 64
Government Relations Advisory Committee, 114

Hahn, Otto, 31

141

health issues, 11, 14–15, 16, 17, 45–52, 59, 82–84, 88–90
Hertzmark, Donald, 117
Hoover, Ruth, 82, 90–91

International Atomic Energy Agency, 89, 111–112
International Commission on Radiological Protection (ICRP), 62
International Red Cross, 89

Jersey Central Power and Light, 42
Johnsrud, Judith, 86

Kamanoff Energy Associates, 125
Kansas Geologic Survey, 113–114
Kemeny Commission, 85–86
Kemeny, John, 84–85

landfills, 100–101
Langer, R. M., 35
Lilienthal, David, 37, 43
linear theory, 61–62
loss-of-coolant accident (LOCA), 25
Lovins, Amory, 117–119, 120–122
Low-level Radioactive Waste Policy Act (LLRWPA), 102–104

Manhattan Project, 32, 34–35
McMahon, Brien, 36
meltdowns, 69, 73, 78, 79, 81–82, 85, 90; at Chernobyl, 12–15
Miller, Bob, 113
Mutual Assured Destruction (MAD), 38

Nagle, Conrad, 104, 105
National Academy of Sciences (NAS), 90, 110
National Council on Radiation Protection and Measurement, 62
National Institutes of Health (NIH), 47

National Reactor Testing Station, 40
National Research Council, 61, 64, 112
natural gas, 10, 30, 53, 54
Nero, Anthony V., 76, 77–78, 79
neutrons, 22, 25, 31, 56–57
New England Coalition on Nuclear Pollution (NECNP), 72, 93
New England Policy Council, 52
nitrous oxide (N2O), 53–54
Norton, Boyd, 31–32, 40
nuclear energy, 21, 42–43, 49, 52, 60, 61
Nuclear Energy Institute (NEI), 18–19, 117–118, 121–123, 125–126; views on disposal, 99, 104, 105–106, 108–109, 110, 111–115; views on pollution, 49, 51, 53–54, 61, 64; views on safety, 67, 69, 71, 83, 85, 89, 91, 92, 96–97
nuclear fuel, 7, 8, 21, 23, 25, 45, 48, 52, 60, 73, 110
Nuclear Industry, 43, 92
Nuclear Information and Resource Service (NIRS), 49
Nuclear Regulatory Commission (NRC), 8, 62, 67–70, 72–75, 84–88, 94–97, 100, 104, 107–108
nuclear waste, 17, 57; disposing, 19, 99–109; high-level, 25, 109–115; low-level, 24, 99–109; storing, 24, 98, 109–115
Nuclear Waste Policy Act, 112

oil, 7, 9, 10, 11, 12, 28, 30, 53–56, 125
operators, 7, 81–82, 96; training of, 8–9, 10, 68–69
Oppenheimer, J. Robert, 32, 33

People's Action for Clean Energy (PACE), 94
Peters, Scott, 18, 19, 51, 96, 115
Pilgrim nuclear plant, 44,

45–48, 51, 52, 57, 60, 69
Pollard, Robert, 73
Prairie Island nuclear plant, 114
Price-Anderson Act, 40–41
protons, 22, 56–57
Putnam, Palmer, 28

radiation, 7–9, 25, 35, 72–73, 81–83, 92, 100; and health issues, 14–15, 16, 38, 45–50, 52, 56–65, 76–79, 88–89; preventing leakage into environment, 25, 67–71
radioactive materials: atoms, 22, 56–57; gases, 45, 47, 57, 64; iodine, 15, 56, 58–60, 81. *See also* fallout; nuclear waste; spent fuel
Rasmussen, Norman, 76
reactors, 18, 21–25, 39–40, 42, 45, 57, 110; boiling water (BWR), 7–9, 16, 23, 39, 45; first, 32; graphite, 12–14; pressurized water (PWR), 23; safety of, 67–70, 76–78, 80–83, 85, 91; at Yankee Rowe, 71–75. *See also* core
reactor vessel, 73–74, 75
Reagan, Ronald, 96
relative biological effectiveness (RBE), 60
renewable energy sources, 118–127.
Resnikoff, Marvin, 100–101, 105, 108–109, 110–111
Reynolds, Albert B., 123–124, 126
Romanyenko, A., 88, 89

Safe Energy Communication Council (SECC), 16–19, 118–119, 121, 122–126; views on disposal, 108, 110, 113, 115; views on pollution, 49–52, 54–56, 61; views on safety, 69, 70–71, 85, 91, 96
"safety-in-depth", 67, 69, 70, 83
Safety Parameter Display Systems, 85

safety systems, 8–9, 12–14, 67–71, 73–74, 75–79; policies regarding, 86–90; public perceptions of, 90–97
Save Chem–Nuclear committee, 102
Scotto, Dan, 124–125
Seaborg, Glen, 27–28
Seabrook nuclear plant, 94, 95–96
Seitz, Frederick, 90, 93
Skinner, Samuel, 114–115
solar power, 16, 50, 118, 120–121
spent fuel, 24–25, 109–115
Strassman, Fritz, 31
sulfur dioxide (SO2), 9, 11, 53–54

Teller, Edward, 80
Three Mile Island nuclear plant, 42, 43, 69, 77, 79, 80–88, 90–91, 94, 95, 117
Truman, Harry S., 36, 37, 39

Union of Concerned Scientists, 11, 72, 73
United Nations, 38
uranium, 21, 34, 54, 57
uranium 235 (U-235), 21–22, 24, 35, 57, 110

Vermont Yankee nuclear plant, 6, 7–10, 16, 39, 46, 68, 69, 93

WASH-1400, 76, 79
water power, 16, 27, 50, 118
Weinberg, Alvin, 126–127
Williams, Carolyn, 102–103
Williams, Hunt, 92–93
wind power, 16, 27, 50, 118
World Health Organization, 15, 89
Worldwatch Institute, 94

Yankee Atomic Electric Company, 71–75, 96
Yankee Rowe. *See* Yankee Atomic Electric Company
Yaroshinskskaya, Alla, 88–89
Yucca Mountains, 98, 111–115

About the Author

Since 1984, **Michael J. Daley** has lived just 18 miles from the Vermont Yankee Nuclear Power Plant featured in the first chapter of this book. Michael remembers visiting Vermont Yankee while in high school and being fascinated with nuclear technology. Then the Three Mile Island accident happened while he was at Dartmouth College. These experiences became the roots of his interest in nuclear power and energy issues. Michael often speaks to students in schools. He always tells them that the use of nuclear power is an open question. Each new generation must come to its own conclusion.

Michael lives in Westminster West, Vermont, with his wife, Jessie Haas. He has published one book on solar energy for young people, and he writes frequently for *Odyssey* magazine.

Photo Acknowledgments

Archive Photos/Lambert, 58; Archive Photos/Russell Reif, 53 (right); Bettmann, 82; Jerry Boucher, 10, 59 (right), 66, 68, 83; Corbis/Bettmann, 48; Robert Del Tredici, 87 (both); © Thomas R. Fletcher, 11 (right); Jerry Hennen, 50; Lyons Daily News, 114; Pheobe Mace, 144; National Archives, 34-35; North Wind Picture Archive, 26; Reuters/Bettmann, 14; Dr. Albert B. Reynolds/University of Virginia, 124; Science VU/Visuals Unlimited, 120; Nancy Smedstad/IPS, 55; UPI/Bettmann, 39; UPI/Bettmann Newsphotos, 88; UPI/Corbis-Bettmann, 17, 33 (both), 37, 41 (both), 53 (left), 80, 95, 119, 123; U.S. Department of Energy, 98, 101, 111; Visuals Unlimited/© SIU, 59 (left); Sally Weigand, 11 (left), 29.